German Two-Seat Fighters of WWI

Jack Herris

German Two-Seat Fighters of WWI

Jack Herris

Great War Aviation Centennial Series #87

Acknowledgements

Brandenburg MLD and KF text courtesy of Colin Owers. Thanks to Colin Owers, Bruno Schmäling, Michael Schmeele, Greg VanWyngarden, and Reinhard Zankl for photographs and information.

Color aircraft profiles © Bob Pearson. Bob notes that while every care is taken, some colors are speculative based on known practices at the time. Purchase Bob's CD of WWI aircraft profiles for $50 US/Canadian, 40 €, or £30, airmail postage included, via Paypal to Bob.

For information on our aviation books, please see our website at: **www.aeronautbooks.com**. Aeronaut Books is looking for photographs of rare German aircraft of WWI for our books. To help please contact the publisher at **jherris@me.com**.

ISBN: 978-1-964637-04-4

Design and layout: Jack Herris
Cover Design: Jack Herris
Color Profiles: Bob Pearson; bpearson@kaien.net
Digital photo editing: Jack Herris

Aeronaut Books

Books for Enthusiasts by Enthusiasts
www.aeronautbooks.com

Table of Contents

Preface

This book is an encyclopedia of German two-seat fighters; all content has been previously published in books on aircraft by manufacturer. Coverage of well-known production types (Halberstadt, Hannover, Roland C.II) in those books is generally in more detail; that is, more photos, more profiles, and scale drawings. Furthermore, this book covers landplanes only; seaplane fighters are covered in *German Seaplane Fighters of WWI*.

Obviously, all airplanes of the CL (light-C) class are included in this book. Howerver, there is often ambiguity between the official designation and the actual design role. As a result, some types designated as C-types but that were intended for the role of light C-type are also included. Examples include the Ago C.V, Albatros C.IX and C.XIII.

Companies such as Halberstadt and Hannover numbered their C and CL types consecutively despite their specific role. For example, Halberstadt used designations CL.II, C.III, CL.IV, C.V, C.VII, and C.VIII, and Hannover used designations CL.II, CL.III, C.IV, and CL.V. Albatros just designated all armed two-seaters as C-types.

Finally, the Roland C.II is also included; it was initially designed as a fast C-class and used as such. A flexible machine gun was provided for the observer, but the inital production batch had no fixed gun for the pilot due to the shortage of synchronizer equipment. After a fixed, synchronized gun was installed for the pilot in later production batches, its crews started to use it as a two-seat fighter due to its speed and agility – without any change in designation.

Introduction

At the dawn of aviation, the envisioned roles were bombing and reconnaissance. It was also realized that, to counter those operations, armed aircraft were required to intercept and destroy the other aircraft.

In the absence of any combat experience, it was not clear what features these interceptors should have. The only experience of combat between moving vessels was naval combat, so a primary proposed solution was aerial cruisers or battleplanes. These aircraft would feature gunners manning flexible machine guns.

In retrospect, synchronized machine guns for the pilot were the most effective solution, and these were patented pre-war in England, France, and Germany. However, these early patents were generally granted for ideas, not working prototypes. Moreover, the idea of aiming a fixed, synchronized gun by aiming the entire airplane was seldom taken seriously, which likely explains the pre-war lack of interest in synchronizers.

Due to lack of machine gun synchronizers, early warplanes, such as the Vickers Gunbus, designed to attack other aircraft were two-seat fighters. In these designs the pilot flew the airplane and a gunner handled the weapon, which was flexible. Typically, the gunner was mounted ahead of the pilot to give him the best possible field of fire, a situation favoring pusher designs.

Combat experience soon demonstrated that the battleplane concept was fundamentally flawed because these airplanes were too slow to intercept hostile aircraft. Many twin-engine battleplanes eventually evolved into their natural role of bombing.

Meanwhile, single-engine aircraft with armed observers forward of the pilot were designed, but only a few, primarily pushers, were placed in operational service. Almost all these were fielded by the Allies; *Idflieg* preferred tractor designs and only Ago and Otto, originally based in Bavaria and not subject to *Idflieg*, produced pushers for the German air service that were placed in operational service.

The Vickers Gunbus and F.E.2 are examples of Allied pusher designs made necessary by lack of synchronizers. After synchronizers because available to the Allies, Britain fielded the Bristol F.2 'Brisfit". Powered by a 275 hp V-12, the F.2 was fast and had good handling. Flown as a standard two-seater, it suffered heavy losses. Once crews started flying them as two-seat fighters, their speed and good handling, coupled with effective armament, made it the most successful Allied two-seat fighter of the war.

For understanding German two-seat fighter development it is helpful to organize the discussion around generations of the fighters, with engine availability being a key feature. Early types were generally created without significant combat experience to guide the requirements. Without clear guidance these can be thought of as the Zero Generation. German types in this generation include:

- AEG C.III (tractor, gunner in front, prototype only)
- Brandenburg MLD, KF (pushers, prototype only)
- Euler combat aircraft (pusher, prototype only)
- Germania DM (tractor, gunner in front, prototype only)
- LVG E.I (monoplane, prototype only)
- Roland C.II

Of the aircraft on this list, the only one that was designed as a fighter from the outset with synchronized gun for the pilot and flexible gun for the gunner was the LVG E.I.

The Roland C.II was a special case. Designed as a standard C-type, its first production batch had a flexible gun for the observer but no synchronized gun for the pilot due to lack of synchronizer availability. Subsequent production batches included the pilot's synchronized gun that it had been designed to carry. When the C.II arrived at the front, it was as fast as operational Allied fighters due to its careful steamlining and use of the new 160 hp Mercedes D.III engine, the first aircraft to reach the front with it. Once the pilot's gun was fitted, its speed and armament enabled its crews to use it as a fighter. Its success encouraged *Idflieg* to establish the CL class (light C class) as a separate catagory. The original concept was the CL class would escort standard C-types and perform gound attack.

First Generation Two-Seat Fighter Designs

These aircraft were the first to be designed to the CL-category requirements, including a synchronized gun for the pilot and flexible machine gun for the gunner. These aircraft were all tractor biplanes that were smaller than standard C-types for better maneuverability.

- AEG C.III, C.VII, C.VIII
- Ago C.IV, C.V
- Albatros C.IX, C.XIII
- Aviatik C.VIII
- Halberstadt CL.II, CL.IV
- Hannover CL.II, CL.III/IIIa

Of these competitors, the AEG, Ago, C.V, Albatros, and Aviatik types remained prototoypes. The Halberstadt and Hannover designs were very successful aircraft. Once the Halberstadt and Hannover CL types arrived at the front, the crews developed new tactics to improve their effectiveness during ground attack and serve as two-seat fighters. Both Halberstadt and Hannover CL types were excellent combat aircraft that were maneuverable and robust enough to dogfight with Allied fighters. Although not as fast as opposing single-seat fighters, they could maneuver with them and the gunner's flexible 'sting' in the tail made them very dangerous in combat.

Second Generation Two-Seat Fighter Designs

These were more advanced designs based on extensive combat experience.

- BFW CL.I, II, III (prototypes only)
- Fokker C.I (production postwar)
- Halberstadt CLS.I (too late for production)
- Hannover CL.V (license production postwar)

Third Generation Two-Seat Fighter Designs

This generation was characterized by technically advanced airframes (monoplane, metal) and/or engines (V8). These included:

- Daimler L8, L14 (V8 power, L14 was a monoplane)
- Junkers CL.I (all metal monoplane)
- Zeppelin-Lindau CL.I, CL.II (all metal biplanes)

Of these designs, only the all-metal Junkers CL.I monoplane reached operational service, albeit in the postwar fighting in the Baltic.

The effectiveness of two-seat fighters in combat was clear to the combatants, and as a result both sides planned to field many more such aircraft had the war lasted into 1919. Britain and France planned to change the composition of their fighter forces to $^{1}/_{3}$ single-seat fighters and $^{2}/_{3}$ two-seat fighters and new types were in development at the Armistice. Germany was also planning enlargement of its two-seat fighter units, but not to the same extent as the Allies.

One change that Germany made at the end of the war was to increase the strength requirements for CL types to make them the same as the requirements for single-seat fighters. The payload was also increased to improve effectiveness in ground attack. As a result, the Halberstadt CLS.I, planned to replace the CL.IV in production, was designed from the beginning to meet fighter strength standards. The pre-production batch of 20 Hannover CL.V aircraft were not built to fighter strength requirements, but subsequent CL.V aircraft were produced to the fighter strength standards. Fourteen were also built in Norway under license postwar and had successful careers.

CL-Class (Two-Seat Fighter) Frontline Inventory

		1914			1915						1916						1917						1918			
Manufacturer and Type		31 Aug	31 Oct	31 Dec	28 Feb	30 Apr	30 Jun	31 Aug	31 Oct	31 Dec	28 Feb	30 Apr	30 Jun	31 Aug	31 Oct	31 Dec	28 Feb	30 Apr	30 Jun	31 Aug	31 Oct	31 Dec	28 Feb	30 Apr	30 Jun	31 Aug
Halberstadt	CL.II																					170	224	342	311	175
	CL.IV																									136
Hannover	CL.II																					162	295	249	72	31
	CL.IIIa																							109	159	233
	CL.III																							51	67	29
Total:																						332	519	751	609	604

Above: The Halberstadt and Hannover CL types reached the front almost simultaneously in late 1917.

Above: Low-level attack by Halberstadt CL.II aircraft.

Above: The Halberstadt CL.IV was a refined development of the CL.II with improved maneuverability. It replaced the CL.II in production and supplemented it at the front. (NMUSAF)

Left: Hannover CL.III powered by the 160 hp Mercedes D.III. One small batch of Hannovers were Mercedes powered; the nearly-identical CL.II and CL.IIIa had to use the 180 hp Argus As.III because the Mercedes was in very high demand for single-seat fighters

AEG C.III

Above: Like the earlier C.I and C.II, the AEG C.III was powered by a 150 hp Benz Bz.III. (The Peter M. Bowers Collection/The Museum of Flight)

The AEG C.III was designed to give the observer the maximum field of fire, achieved by placing the top wing at the lever of the top of the fuselage and the observer in the front cockpit. The Roland C.II to similar format, although with pilot in front, first flew in October 1915 and the AEG C.III, clearly inspired by the Roland, flew two months later.

The C.III was developed from the C.II and used the same 150 hp Benz Bz.III engine. One or two prototypes were built but the insufficient wing gap caused interference in the airflow between the upper and lower wings, limiting performance and flying qualities, and further development was abandoned in favor of more conventional configurations.

Below: The C.III was designed for a more offensive role than the C.I or C.II and the observer with his flexible gun was in the front cockpit, giving him a good field of fire, but it was not synchronized so could not safely fire through the propeller arc.

AEG C.VII

In August 1916 *Idflieg* issued a requirement for a two-seat escort aircraft. This was to be a lightened C-type, later categorized as a CL-type, powered by a 160–180 hp engine. AEG was one of the companies that responded with a proposal, and in October *Idflieg* gave AEG a contract for three prototype C.VII aircraft. The C.VII was powered by a 160 hp Mercedes D.III and the first aircraft was completed in December 1916. It was about 20% smaller than the C.IV powered by the same engine. Initial flight testing was apparently done in January 1917 at Nieder-Neuendorf, and the slightly modified C.VII prototype underwent testing in March. The initial *Idflieg* report (dated April 30, 1917) said the C.VII offered little that was new. The observer's field of fire was considered poor for an escort aircraft and the performance, while superior to the C.IV, was not as good as expected, although flying qualities were acceptable. The C.VII was recommended for service on the less-demanding Eastern Front providing the load tests were passed. That recommendation was not compelling and no production was undertaken.

The third C.VII prototype was built with a highly-swept-back wing with minimum gap, intended to improve the gunner's field of fire. However, the competing Halberstadt and Hannover CL-types were superior and were placed in production while the AEG C.VII was not.

Above & Right: Either the first or second AEG C.VII prototype is shown here. Designed as an escort fighter, a category that later became the CL-class, the C.VII was acceptable but not superior, and the excellent Halberstadt CL.II and Hannover CL.II were produced instead.

Above & Below: The third AEG C.VII prototype featured a swept-back upper wing with minimum gap to improve the gunner's field of fire. However, the competing Halberstadt and Hannover CL-types were still superior aircraft.

AEG C.VIII & C.VIII Triplane

The AEG C.VIII was an attempt by AEG to improve the C.VII for the light two-seater escort role. Three prototypes were ordered, all powered by the 160 hp Mercedes D.III. Apparently two were completed as biplanes, the first in July 1917, and the third as a triplane. It should be remembered that the summer of 1917 was the height of *Idflieg's* triplane craze, which no doubt influenced AEG to complete one C.VIII as a triplane. AEG apparently hoped to achieve high speed by reducing drag to a minimum; the AEG history claims 190 km/h but engineering specifications show a top speed of 170 km/h. If the higher speed were actually achieved the C.VIII might have been ordered into production, but it was not and a reasonable conclusion is the 190 km/h top speed was not reached.

The C.VIII triplane used the engine, fuselage, and tail of the biplane C.VIII. Completed in October 1917 and flight tested in November, the C.VIII triplane climbed to 5000 meters in 34 minutes, a climb equivalent to the Albatros D.V single-seat fighter. With 80 kg less load it climbed to 5000 meters in 24 minutes. However, it was slower than the biplane C.VIII and flight-testing revealed the C.VIII triplane to have unsatisfactory flying qualities. AEG had proposed a smaller, lighter triplane in November but the disappointing flight-test results of the C.VIII triplane eliminated that idea as well.

Above & Following Page, Above: One of the first two AEG C.VIII prototypes; the C.VIII was a cleaner, more aerodynamic design than the C.VII.

Above: The third AEG C.VIII prototype was built as a triplane. The climb rate was good but like most triplanes it was slower than the biplane from which it was derived, and it did not handle as well. (The Peter M. Bowers Collection/The Museum of Flight)

AEG C.IV C.4871/17 of *FAA* 303 based in Palestine

Ago C.IV

Above: Ago C.IV 2747/16 is an early machine in the initial production configuration from the first production batch. Camouflaged in a two color scheme, there is no fixed fin, ailerons are fitted on the upper wing only, and the broad interplane struts are painted the underside colors. The forward inboard interplane struts are omitted to enable the observer a larger field of fire forward and the rectangular wire in front of the observer's cockpit is to prevent the observer firing into the propeller arc. The fixed machine gun for the pilot is mounted over the engine. (Bruno Schmäling)

Ago had exhausted the potential of the pusher configuration with its C.I, C.II, and C.III designs. To achieve better performance and to create a design that could be sold to the Prussian-dominated *Fliegertruppe*, Ago turned to the tractor configuration.

The result was the Ago C.IV, an innovative design of conventional construction that offered much better speed and climb than the pushers that preceded it.

The Ago C.IV was most easily distinguished from other two-seaters in having a sharply-tapered wing planform. Most WWI aircraft had untapered wings of rectangular planform for ease of manufacturer. However, the induced drag (drag induced by lift) depends on the wing planform, and theoretical calculations indicate that an elliptical wing planform minimizes induced drag. (The Spitfire was an example of an actual design using this principle.) An elliptical planform wing is difficult to manufacture. However, it can be approximated by the tapered wing planform, which is a compromise. The tapered wing has less induced drag than a rectangular wing but is harder to build. It has more drag than an elliptical wing but is easier to build.

The resulting C.IV with 220 hp Benz Bz.IV engine was faster than most two-seat reconnaissance airplanes. It was carefully designed to have clean lines and be relatively streamlined for the time. The combination of good power, uncluttered lines, and tapered wing planform gave it good speed and climb for its time. It was much faster than the C.I that Ago had been building, and had the tractor configuration preferred by the *Fliegertruppe*. Ago appeared to have found the formula, and the C.IV was ordered in quantity starting in June 1916.

Actual design of the Ago C.IV had started in late 1915. Another innovative feature of the C.IV was the lack of a forward, inboard bracing strut to provide a better field of fire forward between the wings for the observer. Ago was given Patent No. 299,149 for this innovation on Oct. 7, 1915, as they were designing the C.IV.

The Ago C.IV prototype flew in early 1916 and immediately demonstrated good performance. In trials flown on May 21 the prototype reached 4,000 m altitude in 32 minutes, fast for the time. This encouraged *Idflieg* to order 24 aircraft in June 1916. Contemporary *Idflieg* files referred

Above & Below: The first production version of the Ago C.IV during type-testing at Adlershof in June 1916. Like the prototype, there is no fixed fin and ailerons are fitted on the upper wing only. Furthermore, it has a propeller spinner and revised nose contours. The forward inboard interplane struts normally seen on two-bay aircraft were omitted to enable the observer a larger field of fire forward. The broad interplane struts were painted in the undersurface color, likely light blue. However, unlike the light-doped prototype, this aircraft had the production two-tone camouflage scheme. The scheme was not very distinct and was green and red/brown. The Ago C.IV had a number of innovative features, some of which had to be modified after operational service, and had a significantly better performance the Ago's earlier pusher designs. (Peter M. Grosz Collection/STDB)

Above & Below: Ago C.IV(Schül) 5513/16. The aircraft is in final production configuration including fixed fin, ailerons on all wings connected by an actuating strut, and a diagonal bracing strut added to the outer interplane strut assembly. The fuselage text reads: *Zulässige Belastung bei vollem Tank 520 kg.* (Permissable load at full tank 520 kg) *Leergewicht 1020 kg.* (empty weight 1020 kg). This may indicate training use. (Peter M. Grosz Collection/STDB)

Ago C.IV Aircraft Production			
Order	**Serials**	**Qty**	**Notes**
June 1916	2740–2763/16	24	Ago
Sept. 1916	3714–3723/16	10	Rathgeber (not completed; reduced to spares)
Oct. 1916	5500–5749/16	250	Schütte-Lanz*
Nov. 1916	3600–3623/16	24	Ago
Nov. 1916	8950–8999/16	50	Ago
Feb. 1917	1521–1570/17	50	Ago
May 1917	5575–5674/17	100	Ago. Few if any delivered. Order transferred to LVG C.II(Ago) trainers in Sept. 1917.
* Order reduced to 150 in July 1917 & cancelled in Oct. 1917 after 66 aircraft were delivered.			

to the Ago C.IV as *the so-called light C-type*. It was thus a precursor of the later CL-class.

Type-testing was completed in mid-June. The testing revealed the need to wing strengthening, and after two reinforcements of the metal wing fittings the required strength was achieved.

An order for 10 aircraft was given to the Munich-based Waggonfabrik Josef Rathgeber AG to bring that company into much-needed aircraft production. Initially the order was for the Rathgeber C.I but was later re-designated the Ago C.IV(Rat) when *Idflieg* rationalized aircraft designations. However, these aircraft were never delivered and the parts were reduced to spares.

Spare production capacity was available at Schütte-Lanz, best known for its airship manufacture as a competitor to Zeppelin. By this time airships were losing favor and the superior Zeppelins could fill requirements. None of Schütte-Lanz's own aircraft designs were selected for production and *Idflieg* ordered 250 Ago C.IV(Schül) aircraft from them to utilize their spare production capacity. The size of the order indicated *Idflieg's* interest in the C.IV.

In December 1916 the Ago C.IV arrived at the front. The pilots, who had been used to the stable and reliable Ago C.I, were anticipating the C.IV to combine these qualities with significantly improved performance. However, they were soon disappointed. As soon as the C.IV went into service, a rash of fuselage failures occurred during landing. However, this problem was quickly corrected in January and deliveries continued.

As the crews logged increasing flight time on the C.IV, fatalities due to side-slipping increased alarmingly. The Ago C.IV was grounded at the end of April pending investigation and solution of the flight-control problem.

The original C.IV prototype and production aircraft had ailerons on the upper wing only and did not have a fixed fin. As a result of the accidents the flight controls of the C.IV were extensively redesigned to improve its flight characteristics. Ailerons were fitted to all wings and a fixed vertical fin was added to the tail. In addition, most all the aircraft delivered by Ago after serial 8950/16 and all built by Schütte-Lanz had an additional bracing strut between the outer wing struts. *Idflieg* also tested Ago C.IV 1521/17 fitted with conventional two-bay bracing. The first Schütte-Lanz built C.IV was type-tested May 7–8, 1917, during which the metal wing fittings failed again and had to be reinforced. Cleared for combat, few Schütte-Lanz built C.IVs reached the front.

The tapered wing planform gave good performance but contributed to the flight-control problem and the technology of the time was not adequate to 'tune' the wing's aerodynamics for good stability. Wings want to stall first at their tips, and this tendency is worsened by tapering the planform. This caused the loss of roll control at the critical moment of a stall. With most rectangular planform wings, the designers 'washed out' the angle of incidence at the wingtips to ensure the wing stalled first at the wing root, countering its tendency to stall first at the tips where the ailerons were located, ensuring the pilot retained aileron control during stall recovery.

After the modified C.IV was returned to the front in the May–June 1917 timeframe, the C.IV was moderately successful but was still unpopular with aircrews due to its tarnished reputation. Continuing pilots' resistance and flight-control problems caused *Idflieg* to ground the C.IV again on September 7, 1917 and cancel production.

The Ago C.IV was the last Ago design to be placed in production. It is included in this volume beause it was referred to as a light C-type by *Idflieg* and was a fairly small, fast two-seater with single-bay bracing. However, it was used as a standard C-type during operations, likely due to its bad handling qualities.

Ago C.V

Above: The Ago C.V was a single-bay biplane using N-struts. The observers' cockpit in this version is raised above the level of the pilot's cockpit to provide the gunner a field of fire forward. (Peter M. Grosz Collection/STDB)

The compact Ago C.V two-seater powered by the 160 hp Mercedes D.III was designed as a CL-type despite its designation. The single-bay design used N-struts and the wing planform was nearly rectangular with ailerons on all four wings. The angle of incidence of the wingtips was washed out so the ailerons would stall last, preserving the pilot's roll control after a stall. The C.V represented a major effort to improve flight characteristics after the serious problems with the earlier Ago C.IV.

Not only was the Ago C.V designed for improved flight characteristics, but it was well-streamlined. No production was undertaken due to the excellence of the competing Halberstadt and Hannover designs.

Above: The Ago C.V with modified rear fuselage contours. (Peter M. Grosz Collection/STDB)
Below: Front view of the Ago C.V. (Peter M. Grosz Collection/STDB)

Albatros C.IX

Above: The Albatros C.IX used the technology of the Albatros fighters. It was designed to meet *Idflieg's* specification for a light C-type. Over-shadowed by the competing designs from Halberstadt and Hannover, it did not go into production. Manfred von Richthofen used this example for personal transportation and is seen here in the rear cockpit.

In August 1916 *Idflieg* issued technical specifications for lightened C-type aircraft powered by 160–180 h.p. engines. The light C-type, or CL-type, was intended to be a two-seat escort fighter, short-range reconnaissance aircraft and artillery spotter, and ground-attack aircraft. Two companies, Halberstadt and Hannover, created excellent CL-designs that entered production in late 1917 and served successfully until the armistice. Albatros also designed a couple of aircraft to this requirement, but these did not enter production.

The Albatros C.IX was built in 1917 and appears to have been designed with the CL-specification in mind. The C.IX was a single-bay biplane that was notably smaller than earlier Albatros two-bay C-types. Unsurprisingly, the C.IX used the same semi-monocoque wood fuselage structure as the successful Albatros fighters and was powered by the same 160 hp Mercedes D.III engine. However, the wing design featured two spars for both upper and lower wings with X-struts in addition to parallel struts for robust interplane bracing. The lower wing was straight with rounded tips, while the upper wing was mounted forward with a large cut-out over the pilot for better visibility and access, and was sharply swept-back to maintain the proper center of lift. Although not fitted to the example shown above, the C.IX was designed to mount the standard German two-seater armament of a fixed Spandau machine gun for the pilot and a flexible Parabellum machine gun for the observer.

The Albatros C.IX did no go into production and only three were built. It is mainly known for its association with the Red Baron, Manfred von Richthofen, who briefly used one as a squadron hack; however, one served with *Flieger Abteilung* 12 in 1918.

Above & Below: Designed in response to *Idflieg's* requirement for a light C-type the Albatros C.IX exhibited excellent workmanship. Design of the C.IX was also more innovative than most Albatros aircraft, although it had a streamlined fuselage characteristic of Albatros aircraft of that time. However, the competing Hannover CL.II was a more compact design that offered better speed and climb and, together with the Halberstadt CL.II, was selected for production in place of the shapely C.IX. (Peter M. Bowers Collection/The Museum of Flight)

Comparison of Albatros C.IX & C.XIII to Hannover CL.II			
	Albatros C.IX	Albatros C.XIII	Hannover CL.II
Engine:	160 hp Mercedes D.III	160 hp Mercedes D.III	180 hp Argus As.III
Wing Span	10.4 m (34.1 ft.)	10.0 m (32.8 ft.)	11.95 m (39.2 ft.)
Length	8.22 m (27.0ft.)	7.8 m (25.6 ft.)	7.8 m (25.6 ft.)
Empty Weight	790 kg (1,742 lb.)	700 kg (1,543 lb.)	773 kg (1,704 lb.)
Loaded Weight	1,150 kg (2,535 lb.)	1,060 kg (2,337 lb.)	1,133 kg (2,498 lb.)
Maximum Speed	155 km/h (97 mph)	165 km/h (103 mph)	165 km/h (103 mph)
Climb to 1000 m	5 min.	4 min.	4 min.
Climb to 4000 m	30 min.	–	–
Climb to 5000 m	–	47 min.	32 min.
Armament:	1 flexible machine gun & 1 fixed machine gun	1 flexible machine gun & 1 fixed machine gun	1 flexible machine gun & 1 fixed machine gun

Above: The excellent Hannover CL.II was placed in production. It had better speed and climb than the Albatros C.IX and was also noted for its maneuverability, something the C.IX was unlikely to have matched. The Albatros C.XIII fared better in comparison but had a single-spar lower wing that caused structural problems in Albatros fighters.

Albatros C.IX 4508/16 serving as a squadron 'hack' in *JG* I and used by Manfred von Richthofen in September/October 1917.

Albatros C.IX prototype mid 1917.

Above: Albatros C.IX in center of a lineup, probably photographed at *Flieger Abteilung* 12 in February 1918. (Peter M. Grosz Collection/SDTB)

Below: Camouflaged Albatros C.IX in front of a hangar, probably photographed at *Flieger Abteilung* 12 in February 1918. (Peter M. Grosz Collection/SDTB)

Above: Albatros C.IX photographed at *Flieger Abteilung* 12 in February 1918. *Idflieg* normally ordered three prototypes of experimental aircraft, one for flight testing, one for structural testing, and a spare. The structural test airframe was normally tested to destruction, leaving two flyable airframes. Manfred von Richthofen used one as a transport, and here is the other. Being continually short of aircraft the German Navy often assigned prototypes to operational service; that was much less common for the Army. Regardless, here is a C.IX prototype at an operational unit – and in night camouflage! Was it used for light bombing? Night fighting? Night reconnaissance? Or all those roles? (Peter M. Grosz Collection/SDTB)

Albatros C.IX photographed at *Flieger Abteilung* 12 in February 1918 with lozenge night camouflage on the fuselage and hexagonal night camouflage on the wings and horizontal tail.

Above: Colorized view of Richthofen in Albatros C.IX 4508/16 at Hannover, Germany, ca. Sept/Oct, 1917.

Below: Richthofen being assisted into his flight gear prior to boarding Albatros C.IX 4508/16. Note he is wearing the localized, strapped-on bandage.

Above: Richthofen in the front cockpit of Albatros C.IX 4508/16 at Hannover, Germany, ca. Sept/Oct, 1917.

Below: Richthofen in the front cockpit of 4508/16, preparing for flight with Mohnicke. The Albatros C.IX had a fairing before the front cockpit that blocked the wind, the clear panels of which allowed light to enter the cockpit.

Above: Lothar von Richthofen (right) and his future wife, Countess Doris von Keyserlingk, sit in the cockpits of Manfred von Richthofen's red-painted Albatros CIX. The Countess appears to be sceptical of flying.

Below: Manfred von Richthofen and Mohnicke in C.IX 4508/16, preparing to take off from Hannover after Richthofen's visit.

Above: Albatros C.IX 4508/16 taking off from Hannover after Manfred von Richthofen's visit.

Below: The nose of Albatros C.IX 4508/16 shows at left in this photo taken in a post-war museum.

Above: Albatros C.IX 4508/16 in flight. Although not confirmed, Richthofen may be at the controls. Note the marked sweepback of the upper wings.

Facing page: Albatros C.IX 4508/16 photographed during Manfred von Richthofen's stop-over at Hannover.

Above: Albatros C.IX 4508/16 in a post-war museum. Legends painted on the fuselage state *Reisenflugzeug des Rittm. Manfred Freiherr v Richthofen* ("Travel plane of *Rittmeister* Manfred *Freiherr* von Richthofen") and *Kriegsflugzeug für Fernaufklärung Alb. CIX 1916/17* ("War plane for reconnaissance Alb. CIX [the years of] 1916/17"). (Courtesy of the DEHLA Collection)

Above: Richthofen's Albatros C.IX 4508/16 on display in the 1930 at the *Deutsche Luftfahrt Sammlung* without tires. This view emphasizes its compact dimensions.

Halberstadt CL.II '6' of *Schlasta* 5

Albatros C.XIII

Above: Colorized photo of the Albatros C.XIII prototype. The streamlined C.XIII was obviously derived from the Albatros D.V fighter. It was much lighter and more compact than the earlier C.IX and was a more convincing design. Although its performance matched the competing Hannover CL.II, it may not have matched the Hannover's excellent maneuverability. Moreover, limited by its single-spar lower wing that had already suffered structural failures in the Albatros D.V, the C.XIII clearly lacked the Hannover's robustness, an essential quality for a successful combat aircraft.

After the failure of the Albators C.IX to reach production for the light C-type role, Albatros created another design to the same *Idflieg* specification. The resulting C.XIII had a single-bay configuration like the C.IX and, obviously based on the Albatros D.V fighter, was intended for the CL-role. A compact, streamlined aircraft, it was powered by the same 160 hp Mercedes D.III used in the D.V fighter and earlier C.IX prototype. The cockpits were very close for crew coordination. Unfortunately, the C.XIII retained the single-spar lower wing already proved structurally unsatisfactory in the Albatros D.V. Armament was the standard German two-seater fit of a fixed Spandau machine gun for the pilot and a flexible Parabellum machine gun for the observer.

Normally *Idflieg* would order three examples of a new prototype. However, only a single C.XIII was apparently built.

Albatros' second attempt at a light C-type, the C.XIII, was more convincing than its C.IX. It was more refined, with no protrusions, and was much lighter than the C.IX or Hannover. It looked aggressive, and, except for high-altitude climb rate, its performance matched the Hannover. However, to achieve its light weight the C.XIII retained the single-spar lower wing of the D.V, so undoubtedly inherited its structural weakness, whereas the Hannover was a very robust design. Furthermore, the C.XIII may not have matched the Hannover's excellent maneuverability. With a more robust two-spar lower wing like that of the Pfalz D.III, the C.XIII might have been competitive. The workmanship of the wood fuselage is impressive.

Aviatik C.VIII

Above: The Aviatik C.VIII was a small, single-bay biplane apparently designed as a light C-type. The C.VIII remained a prototype; the Halberstadt and Hannover designs were produced in quantity instead and both were very successful.

The Aviatik C.VII was apparently an un-built project, making the 1917 C.VIII the next Aviatik two-seater design actually built. The C.VIII was a small, single-bay biplane apparently designed as a light C-type in competition with the Halberstadt CL.II and Hannover CL.II. Like these competitors the C.VIII was powered by a 160 hp Mercedes D.III. To give sufficient gap between the wings to avoid aerodynamic interference, the lower wing was attached to a keel below the fuselage and braced with additional struts. Although no performance data for the C.VIII servive, it remained a prototype.

Halberstadt CL.II '6' of *Schlasta* 6

Halberstadt CL.II '2' of *Schlasta* 33

Halberstadt CL.II 14258/17
Seefrosta
March 1918

Halberstadt CL.II '5' *Thea* of *Schlasta* 27

BFW Aircraft

BFW (Bayerische Flugzeugwerke AG) was created in Munich February 1916 from the unprofitable aircraft builder Otto-Flugzeugwerke.

Due to the need for immediate aircraft production for the ongoing war, initially there was no time for development work and BFW manufactured aircraft under license from Albatros Flugzeugwerke. Within a month of being set up, the company was able to supply aircraft to the war ministries of Prussia and Bavaria. However, from the beginning major quality problems were encountered, a legacy of the preceding Otto works. The German aircrews frequently complained about the serious defects that appeared in the first machines from BFW. The same thing had happened with the aircraft from the predecessor company run by Gustav Otto. It required significant organizational changes and more intensive supervision of the assembly line to succeed in resolving these problems by the end of 1916. BFW then started producing more than 200 aircraft a month, with their workforce growing to 3,000 and becoming the largest aircraft manufacturers in Bavaria after the Pfalz company. Types BFW built under license included the Albatros B.II, C.III trainers, Albatros C.XII, and the Halberstadt CL.II and C.V

BFW CL.I

Above: BFW CL.I prototype powered by a 160 hp Mercedes D.III engine. (Peter M. Grosz Collection/STDB)

Although BFW was in the business of repairing airplanes and building others, mostly intended as trainers, under license, they did build several original designs. The purpose for this has been lost to time. Were the BFW designs intended to let engineers burn off their creative energy or did they seriously want to create a new design team? BFW had the advantage of building the Halberstadt CL.II to learn from. The CL.I was a small biplane with interesting circular frontal radiator. The interplane struts were I-struts. The sole prototype built in 1916/1917 achieved a speed of 175 km/h.

Above: The delicate look of the struts on the BFW CL.I prototype indicates it may not have been sturdy enough for the demanding two-seat fighter role of CL-type aircraft. (Peter M. Grosz Collection/STDB)

BFW CL.II

Above: BFW CL.II prototype powered by a 185 hp Mana III engine. (Peter M. Grosz Collection/STDB)

Above & Below: The BFW CL.I and CL.II appear to have used the same airframe design. (Peter M. Grosz Collection/STDB)

The BFW CL.II was powered by a 185 hp Mana III engine and flew in July 1918. The aircraft appeared to have essentially the same airframe as the earlier CL.I.

No documentation remains to say why the BFW company decided to design original aircraft in the first place or why the designs were of the CL type. It makes sense BFW would avoid multi-engine types due to low demand, and avoid floatplanes for the same reason. Fighters were especially competitive, so perhaps C-types, CL-types, and N-types would come up as candidates, and BFW built C-types and the Halberstadt CL.II under license so had some familiarity with them. CL-types required robust, maneuverable aircraft of good performance. In contrast, the BFW CL designs all seemed somewhat delicate with only one interplane strut. Furthermore, the incumbent producers, Hannover and Halberstadt, were already working on their next-generation designs. BFW did not have a realistic chance of success as a new, untried design team building CL-type aircraft in competition with Hannover and Halbertadt who were already building very successful CL-types in quantity.

BFW CL.III

Above & Below: BFW CL.III prototype powered by a 200 hp Benz Bz.IV engine. Two-seaters with that engine were normally classified as C-types. (Peter M. Grosz Collection/STDB)

The BFW CL.III, powered by a 200 hp Benz Bz.IV engine, appeared to use the same basic airframe as the BFW CL.I and CL.II. Like them, it remained a sole prototype. Use of a 200 hp engine normally put an aircraft in the C-class, not CL-class, although references to it call it the CL.III. Certainly its airframe was CL-class.

During flight testing it proved to be faster than the Albatros D.V. The BFW CL.IV was similar to the CL.III but remained a project.

Above & Below: The BFW CL.I, CL.II, and CL.III all appear to have the same airframe. (Peter M. Grosz Collection/STDB)

BFW Monoplane

Above: BFW monoplane prototype powered by a 160 hp Mercedes D.III engine.

In addition to the three CL-prototypes, BFW also built an experimental monoplane. The reason this airplane was built is not clear but testing the eccentric wing design may be the reason. The problem this design was intended to solve is not known. The prototype used a fuselage and tail based on the previous CL-prototypes coupled with a shoulder-mounted monoplane with forward sweep and unusual planform. The engine was a 160 hp Mercedes D.III engine.

The wing was of thin airfoil, limiting its strength.

Built in 1918, the aircraft remained a single prototype.

Halberstadt CL.II of *Schlasta* 12

Brandenburg Type MLD

Above: The Type MLD was a typical early pusher design with the tail surfaces supported by twin booms.

The MLD was a 1915 pusher design with twin booms carrying the tail surfaces that incorporated triple rudders. It carried a two-man crew and was equipped with a machine gun. Designed before the arrival of the first true fighters, its intended role was probably as much reconnaissance as fighting. It was not successful and it is not certain for whom the aircraft was intended.
Text courtesy Colin Owers.

Type MLD Specifications

Source	Typenschau	Brandenburg 3-View
Span, m	16.50/15.70	16.500/15.700
Length, m	10.30	10.300
Wing Area	55.80 m^2	
Motor	150-hp Benz	150-hp Benz

Halberstadt CL.IV '6' of *Schlasta* 6

Brandenburg Type KF

Above & Below: The first Brandenburg that could be considered a true fighter. The KF had the gunner seated in front of the pilot and the fixed spigot mount for the machine gun may be seen in the photograph. The ply tail booms are well shown.

The KF was Brandenburg's first attempt to design a dedicated fighter although the machine was intended to act in the reconnaissance role as well. Appearing early in 1916 the KF was a pusher with the observer/gunner seated in front of the pilot. The tail surfaces were carried by twin booms that came from the lower wings. The machine was of conventional Brandenburg mixed wood and fabric construction with a ply covered nacelle and tail booms. Work was discontinued as it offered no improvement over types already in production.*
Text courtesy Colin Owers.

Type KF Specifications			
Source	**Typen-schau**	**Green & Swanborough**	**Brandenburg 3-View**
Span, m	11.60/ 10.80	11.60	11.600/ 10.800
Length, m	8.48	8.48	8.480
Wing Area	38.60 m^2	38.6 m^2	38.6 m^2
Empty Wt.	760 kg	760 kg	760 kg
Loaded Wt.	1,290 kg	1,290 kg	1,290 kg
Speed	—	140 km/h	—
Motor	—	150-hp Benz Bz.III	150-hp Benz

* The Type KF is illustrated in the *Typenschau* by a three-view drawing.

Daimler L 8

This Page: The Daimler L8 (aka CL.I) two-seat fighter prototype had a conventional, single-bay wing and slab-sided fuselage. Flying surfaces were covered in printed camouflage fabric. It was powered by the experimental Mercedes D.IIIb V-8 engine. Its V-8 engine finally went into production, but unfortunately too late to power any fighters in combat.

The Daimler L.8 was a single-bay biplane powered by a Mercedes D.IIIbm engine. Essentially it was an enlarged D.I. Armament was one fixed machine gun for the pilot and a flexible gun for the observer. Photographic equipment was carried, but no wireless equipment was fitted.

Daimler L.8 (aka CL.I) Specifications		
Engine:	185 hp Mercedes D.IIIb V-8	
Wing:	Span	11.820 m
General:	Length	7.465 m

Daimler L.14

Above & Below: The Daimler L14 two-seat fighter was a two-seat derivative of the high-performance L11 single-seat fighter. Powered by Daimler's 185 hp Mercedes D.IIIbm geared V-8, the L14 was the final Daimler aircraft to be built and flown. Unfortunately, no performance data is available. Its close resemblance to the L.11 is obvious.

The Daimler L 14, designed by Klemm, was a slightly larger, two-seat fighter derived from the L 11 fighter. First flown by von Thüna in the fall of 1919, had it been tested by *Idflieg* it would have probably been designated the CL.II. It was the final Daimler design. It was offered to Chile without result. Unfortunately, no performance data is available.

Daimler L 14 (aka CL.II) Specifications

Engine:	185 hp Mercedes D.IIIbm V-8	
Wing	Span	12.950 m
General	Length	7.280 m

Above, Below, & Bottom: The Daimler L14 two-seat fighter prototype. (Peter M. Grosz collection, STDB)

Euler

Above: Euler combat airplane Work Number 188 with 2 machine-guns and no nose wheel.

In the autumn of 1915 Euler built two new pusher biplanes, Work Numbers 188 and 189, as two-seat fighters. Both were powered by the then-new 160 hp Mercedes D.III engine, had an empty weight of 600 kg, and had 48 m² wing area. The framework of the nacelle was metal. These types appear to have had identical wings and tail but varied in the design of the crew nacelle.

Work Number 188 mounted two machine guns, one for each crewmember. The gunner sat in the rear cockpit that was raised above the pilot and had a flexible machine gun with a 360° field of fire above the upper wing. The pilot sat in the front cockpit and his gun was fixed. The aircraft had a simple undercarriage without nose wheel. Phots show that the observer's position had alternate designs, one fully faired and the other was more open.

This aircraft crashed on 11 November 1915 and was destroyed.

Work Number 189 had a single flexible machine gun for the gunner who occupied the front cockpit and had a nose wheel. The nose wheel was half-submerged in the fuselage and was apparently fitted to handle nose-overs in case of bad landings.

It was tranferred to Döberitz on 29 November 1915 for evaluation. Flight testing revealed unsatisfactory handling characteristics. No production followed; both types remained single prototypes.

Above: New MG arrangement for Work Number 188.

Below: Construction details for Work Number 189. Note the metal frame.

Above & Below: Armament for pilot and gunner in Work Number 188; different observer designs are shown.

Above: Equipment variation of Work Number 188 with 2 machine guns.

Engine installation viewed from the rear.

Above: Euler Word Number 189 with movable MG in front with bracket after Euler's patent.

Above: Euler combat airplane Work Number 189 with nose wheel and 1 flexible MG.

Fokker V38 /C.I

Above & Below: The Fokker V38 was an enlarged, two-seat derivative of the BMW-powered D.VII. It was the prototype of the Fokker C.I that was produced in series postwar. (Below Peter M. Grosz Collection/STDB)

The V38, W/N 3658, was an enlarged two-seat derivative of the D.VII with the 185 hp BMW IIIa. It was the prototype of the Fokker C.I, which was only built in series after the war. Armament was one synchronized gun for the pilot and a flexible gun for the observer.

Above: The Fokker V38 was an enlarged, two-seat derivative of the BMW-powered D.VII. It was the prototype of the Fokker C.I that was produced in series postwar. (Peter M. Grosz Collection/STDB)

Above: The uncovered fuselage and tail of the Fokker V38 prototype of the Fokker C.I showing its steel-tube structure that was standard for Fokker.

Above & Below: The Fokker V38 was an enlarged, two-seat derivative of the BMW-powered D.VII. It was the prototype of the Fokker C.I that was produced in series in Holland postwar. The longer wing span compared to the D.VII is evident. (Peter M. Grosz Collection/STDB)

Above & Below: Fokker C.I; these photographs were used in Fokker advertisements for the C.I.

Above: An early LVA Fokker C.I with the orange circle insignia.

Above, Below, & Facing Page, Below: Three views of LVA Fokker C.I No. 485 with the first Netherlands national insignia. Note the bombs carried under the fuselage.

Germania DM Two-Seat Fighter

Above: Germania DM two-seat fighter prototype of 1915 had the gunner in the front cockpit.

The "DB", a larger two-seat version of the Germania JM fighter was developed in parallel with the JM. It was designed as a two-seater fighter, with a 180 hp Argus As III engine. The observer/gunner had a free field of fire in front over the propeller arc. As with the other Germania-designed aircraft, only one DM prototype was built and tested in 1915.

The DB two-seat fighter utilized the efficient *Walfisch* (Whale) fuselage configuration of the JM model which endowed the gunner, who sat ahead of the pilot, with a broad forward field of fire. A two-bay biplane. the DB was powered by a water-cooled engine, carried a single flexible machine gun on a ring mounting in the forward cockpit, and underwent trials in September 1915.

The DB underwent an intensive technical investigation through the Royal Marine *Marine-Landflugzeug-Abnahme-Kommission* (Marine Landplane Inspection Commission) situated in Berlin-Johannisthal.

Although the general recommendation of the investigation team supported the introduction of this unique type, a production order was never signed. Egwin Leiber felt that the effort required to implement all the required technical changes was too enormous. Therefore, Germania Flugzeugwerke decided to stop any further development of this type.

This was likely a good decision because the propeller arc blocked part of the most important field of fire, and the competing Roland C.II was better armed (fixed gun for the pilot in front cockpit, flexible gun for observer) and faster due to its better streamlining. The DM greatly resembled the AEG C.III; both used the *Walfisch* streamlining and had the same basic configuration. In contrast, Roland engineers were more creative; they applied the *Walfisch* streamlining in three dimensions instead of side view only. Instead of the rectangular cross section of the AEG C.III and Germania DM, the Roland used a lower-drag oval cross-section. In addition, Roland had to develop the complex *Wickelrumpf* construction method to build it. Made of readily-available wood, the method was strong, streamlined, but labor-intensive due to its complexity. The resulting Roland C.II was very successful and well-known to historians; the AEG C.III and Germania DM were failed designs that remained sole prototypes and are largely forgotten today.

Right: Germania DM two-seat fighter prototype in its hangar.

Halberstadt CL.II '4' of *Schlasta* 23

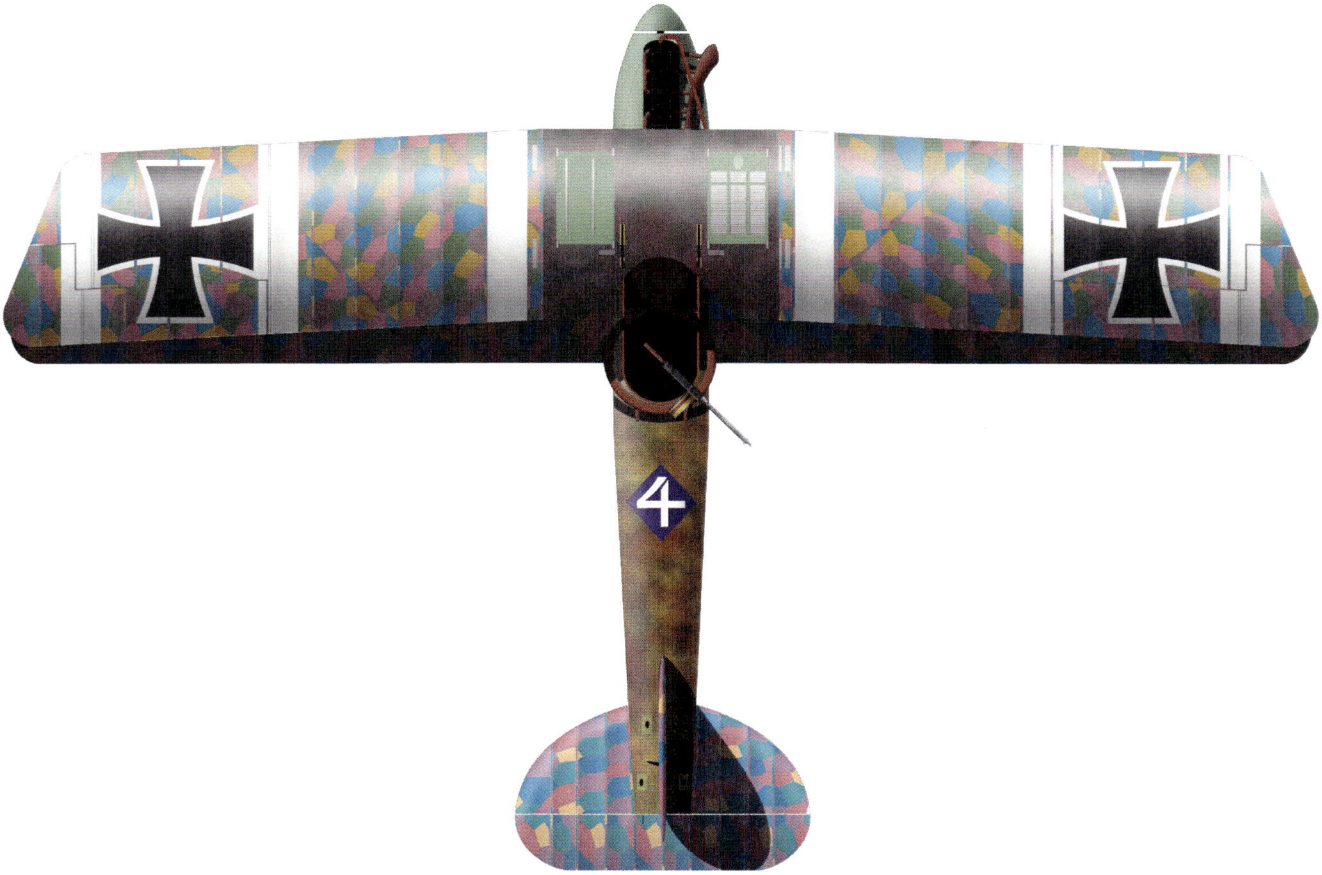

Halberstadt CL.II

Above: Halberstadt CL.II prototype. (Peter M. Grosz collection/STDB)

In August 1916 *Idflieg* issued a specification for a new, light C-type, powered by a 160–180 hp engine. The new CL-type was intended to be a two-seat escort fighter to protect regular C-types from attack. Because the new type was to be a two-seat fighter, *Idflieg* required the aircraft to be 250–260 kg lighter than the standard C-types.

A number of German manufacturers responded to *Idflieg's* requirement; however, only Hannover and Halberstadt responded with successful designs that went into production. Initially the designation Halberstadt C.I was given to the DFW C.V(Halb), the Halberstadt-built DFW C.V, so Halberstadt's successful design was the CL.II. Later, an improved development, the Halberstadt CL.IV, reached the front.

Idflieg ordered three CL-class prototypes from Halberstadt in November 1916. These were the Halberstadt C.II 9901–9903/16, all powered by the 160 hp Mercedes D.III, an engine widely used by German fighters. The resulting C.II (later designated CL.II to reflect its category) was rolled out in April 1917. The prototype had many features first noted in the Halberstadt D.IV. Type-tested over 2–7 May 1917, the load tests were successful but then *Idflieg* required the wing span to be increased, probably to improve climb and ceiling. Due to the increased wing span the load tests had to be repeated later in May and the revised prototype passed.

Based on the good test results, *Idflieg* ordered 100 Halberstadt CL.II aircraft. Eventually 900 CL.II fighters were ordered from Halberstadt and 400 from BFW.

Halberstadt CL.II C.5675/17 was test flown at HFW on July 13, 1917, and Halberstadt CL.II C.5717/17 was assigned to a unit on September 25, 1917. This probably shows that the Halberstadt CL.II first reached the front beginning in September. However, the *Frontbeststand* tables for the C types and CL types do not reflect that.

They were supplied to *Schutzstaffeln* (*Schustas* –protection flights) and reconnaissance *Flieger Abteilungs* as two-seat fighter escorts. The CL.II was immediately successful and quickly earned a good reputation. It was fast, robust, maneuverable, and had a good rate of climb. As important, it had safe flight characteristics and the gunner had an excellent field of fire and field of view. However, it could not keep up with the 'higher-horsepower' two-seat C-types (perhaps the 260 hp Rumpler C.IV?). The undercarriage was considered 'somewhat weak' and liable to collapse unless the landing was very good. However, this was quickly fixed.

The CL.II was compact and therefore the gunner's cockpit was criticized as being too cramped. The CL.II had mountings for a wireless and camera; however, the CL.II was primarily used as a two-seat fighter and seldom carried them.

In the fall of 1917 *Flieger Abteilung (A)* 255 reported that the CL.II was a very effective ground-attack aircraft due to its maneuverability, enabling it to avoid ground-fire. Furthermore, due to its maneuverability and gunner, it could also successfully engage Allied fighters.

The CL.II was assigned to several fighter units for evaluation with operating together with single-seat fighters. After flying the CL.II with a gunner, *Lt.* Kroll of *Jasta* 24 felt the very 'maneuverable' CL.II was on a speed and climb performance level with the Albatros D.III and D.V. CL.II 5717/17 was tested (on Oct. 30, 1917) with twin Parabellum guns for the gunner; the additional weight reduced performance to below that needed to operate effectively with the Albatros fighters. The CL.II was also tested as a night fighter.

Idflieg reported in April 1918 that the Halberstadt CL.II was preferred to all other types by the *Schlactstaffeln*. CL.II losses were very low and it was greatly respected by Allied aircraft.

The CL.II continued in full operational service until the Armistice. It was joined in July 1918 by the improved Halberstadt CL.IV which was even more maneuverable that the CL.II.

The British evaluated captured CL.II 15342/17 brought down on 9 June 1918. Their evaluation was very positive for a German aircraft:

"The Halberstadt (CL.II) represents, in all probability, the high water mark of two-seater German aeroplane construction, as it is not only well and strongly constructed but its general behavior in the air is good according to modern fighting standard.

"Pilots report the machine light and comfortable to fly. The maneuverability is good and this feature taken in conjunction with an exceptionally fine view of the pilot and observer and the field of fire of the latter, makes the machine one to be reckoned with as a 'two-seat fighter,' although the climb and speed performances are poor judged by contemporary British standards."

This last comment is indication the CL.II needed more power, such as could have been provided by a V-8.

There is little information available on two developments of the CL.II with more powerful engines. BFW was producing the CL.IIa powered by the 180 hp Argus As.IIIa engine by late 1918 and a few may have reached the front. Eleven examples were discovered at the BFW factory after the war by the Inter-allied Armistice Commission. Furthermore, the Polish Air Service flew some in 1919–1920.

Another version of the CL.II flew with the superior 185 hp BMW.IIIa engine in January 1918. No production ensured because all the BMW engines were required for fighters and none were available for CL-types. The BMW engine in particular would address the British criticism of its poor speed and climb, and in fact the BMW was installed in the advanced Hannover CL.V while high-altitude 180 hp Mercedes D.IIIaü engine, comparable to the BMW, was installed for the advanced Halberstadt CLS.I that was an even smaller, lighter, and more maneuverable development of the Halberstadt CL.IV.

Additional tests were done with the CL.II. In one, a four-bladed propeller was tested. Another involved the fitting of twin-fixed machine guns for the pilot. Finally, another CL.II was used as a test-bed for early supercharger experiments.

Above: Halberstadt CL.II prototype. (Peter M. Grosz collection/STDB)

Above: Halberstadt CL.II '4' of *Schlasta* 2 photographed with its crew in 1918. The black tailplane and striped struts are the unit markings. (Peter M. Grosz collection/STDB)

Halberstadt CL-Type Specifications			
Spec \ Type	**CL.II Prototype**	**CL.II Production**	**CL.IV**
Engine	160 hp Mercedes D.III	160 hp Mercedes D.III	170 hp Mercedes D.IIIa
Span, upper	9.40 m	10.77 m	10.70 m
Span, lower	9.40 m	10.65 m	9.98 m
Chord, upper	1.6 m	1.60 m	1.60 m
Chord, lower	1.30 m	1.30 m	1.30 m
Wing Area	24.0 m^2	27.5 m^2	26.66 m^2
Length	7.32 m	7.30 m	6.50 m (later 6.89 m)
Height	–	–	2.70 m
Weight empty	701 kg	773 kg	700 kg
Weight loaded	1,071 kg	1,133 kg	1,040 kg
Speed	165 km/h	165 km/h	165 km/h
Climb			
1,000 minutes	–	3.5	4.3
2,000 minutes	–	–	5.5
3,000 minutes	–	–	12.7
4,000 minutes	–	–	22.1
5,000 minutes	–	38	30.8
CL.IV(Rol) Empty Wt. 720 kg, Loaded Wt. 1,060 kg.			

Halberstadt CL.II Production Orders			
Date	**Quantity**	**Serials**	**Notes**
November 1916	3	9902–9904/16	Prototypes. C.9903/16 was in operational use with *Schusta* 22b from September to December 1917.
May 1917	100	5675–5774/17	
June 1917	100	6300–6399/17	
October 1917	200	14200–14399/17	
November 1917	200	15300–15499/17	
February 1918	100	700–799/18	BFW-built
March 1918	100	1161–1260/18	
March 1918	(200)	?	Reinhard Zankl thinks this March 1918 order for 200 aircraft is the C.2800–2999/18 order for 200. But after the first 100 aircraft C.2800–2899/18, the rest were changed to Halb C.V(Bay) 2900–2999/18.
April 1918	100	2800–2999/18	CL.IIa(BFW) with Argus As.IIIa engine
3 prototypes normally ordered, one for flight tests, one for static tests, and one spare.			

Above & Below: Additional views of the Halberstadt CL.II prototypes. The below photo shows the machine is now covered on top and sides in two-tone camouflage, very probably light and dark green. (Peter M. Grosz collection/STDB)

Above & Below: Halberstadt CL.II prototype photographed in May 1917. (Peter M. Grosz collection/STDB)

Above & Below: Additional views of the Halberstadt CL.II prototype. The upper wing is in one piece. The nose is carefully streamlined; the production aircraft did not fully cowl the engine as seen on this prototype. The clever design gave the observer an excellent field of fire. The occupants were seated in close proximity to facilitate communication during combat and both have an excellent field of view. (Peter M. Grosz collection/STDB)

Above & Below: Additional views of the Halberstadt CL.II prototype. (Peter M. Grosz collection/STDB)

Above & Below: An early Halberstadt CL.II visiting the Fokker airfeld at Schwerin-Görries. A Fokker employee is holding a meter stick on the fuselage so it can be photographed in scale. There is a lack of factory-applied identification markings that was typical of Halberstadt-built aircraft. This machine had a typical finish for early batches of Halberstadt factory camouflage. The fabric on the sides and top was five-color light printed pattern (the lozenge fabric production from Augsburger Kattun Fabric in Augsburg (Augsburg Cotton factory) was started from that light pattern so many factories start covering aeroplanes from that light pattern, then used normal pattern on the undersides). The fuselage speckled camouflage was applied over large lozenges from colors similar to printed lozenge light pattern. The bottom of the fuselage remained in light yellow primer. Similarly the bottom part of wings and tail were covered in unpainted, clear lacquered fabric. (Peter M. Grosz collection/STDB)

Above & Below: An early Halberstadt CL.II wearing the Halberstadt-factory colors of the time. The undersurfaces of the wings, tail, and fuselage were not camouflaged in dark colors. (Peter M. Grosz collection/STDB)

Halberstadt CL.II '2' *Brünhilde* of *Schlasta* 27b

Above: A ground crewman loads grenades on Halberstadt CL.II '2' *Brünhilde* of *Schlactstaffel* 27. (Peter M. Grosz collection/STDB)

Facing Page: Crewmen loading weapons on a Halberstadt CL.II of *Schlactstaffel* 27. (Peter M. Grosz collection/STDB)

Below: The ground crew of *Schlactstaffel* 27 prepare grenades and fragmentation bombs before loading them on the unit's Halberstadt CL.II aircraft. The aircraft, in 1918 insignia, have unit insignia and individual names. (Peter M. Grosz collection/STDB)

3030.

Above: Aircrew in front of their Halberstadt CL.II of *Schlasta* 26b in 1918 insignia. (Peter M. Grosz collection/STDB)

Below: Pilot *Unteroffizier* Fritz Kuhlmann of *Schlasta* 26b looks on as a ground crewman repairs the wing fabric of his Halberstadt CL.II '6'. The insignia are 1917 style. Note the flamboyant unit markings. (Peter M. Grosz collection/STDB)

Above: A crew of *Schlasta* 26b is ready for their next mission in their Halberstadt CL.II. The wire guard on the upper wing is there to prevent the gunner from damaging the propeller by firing through the propeller arc. (Peter M. Grosz collection/STDB)

Above: A Halberstadt CL.II of *Schlasta* 26b is waiting for its next mission. (Peter M. Grosz collection/STDB)

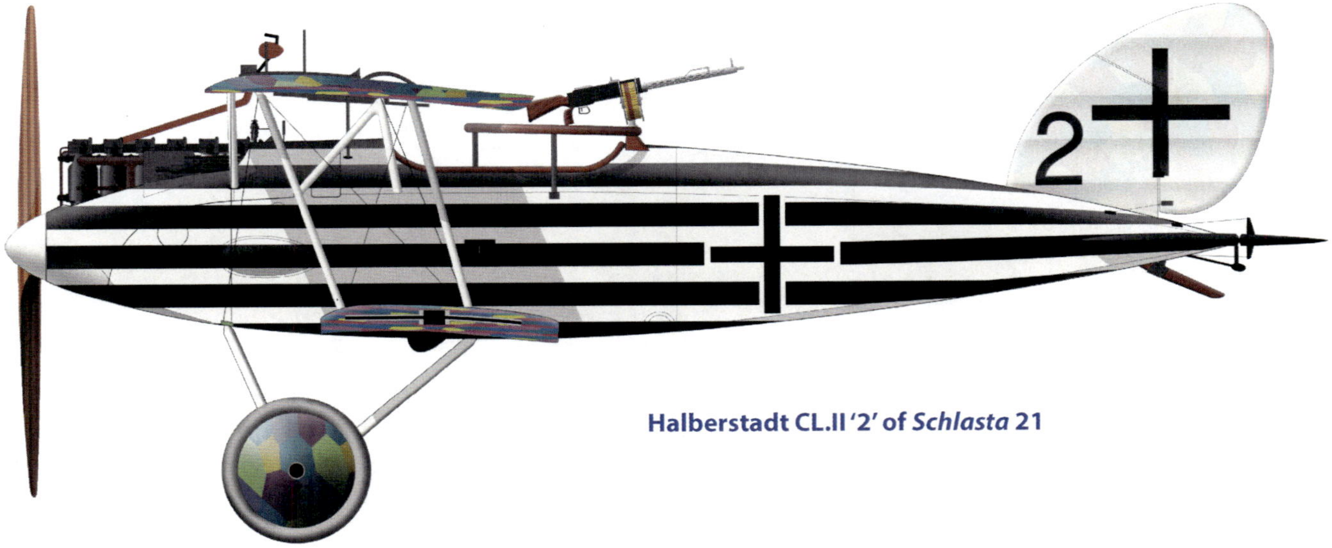

Halberstadt CL.II '2' of *Schlasta* 21

Halberstadt CL.II '5' of *Schlasta* 24

Halberstadt CL.II '4' of *Schlasta* 25

Halberstadt CL.II '4' of *Schlasta* 26

Halberstadt CL.IV

Above: Restored Halberstadt CL.IV of the commanding officer of *Schlachtstaffel* 21. This aircraft is in the National Museum of the USAF. The compact design of the Halberstadt CL.IV is evident.

Developed from the highly successful Halberstadt CL.II, the CL.IV had a shorter fuselage and lighter weight due to careful attention to detail design and the reduced fuselage size. Powered by the same 170 hp Mercedes D.IIIa inline engine as used in later production CL.II aircraft, the empty aircraft weighed 161 pounds (73 kg) less than the CL.II. Aerodynamics were improved with better streamlining and a new wing section. The Mercedes D.IIIa was generally referred to as an 160 hp engine like the D.III, but had slightly more power due to its refinements and actually produced 165–170 hp.

The CL.IV first flew in February 1918. It immediately demonstrated excellent climb and maneuverability. The type test was performed between 5–27 March. The load tests took three weeks because the designer, Karl Theis, had submitted three different sets of wings each with a slightly different structure. The strongest wings were chosen for production.

In April–May *Idflieg* assigned the first production order for 150 of the Halberstadt CL.IV. An engine shortage delayed production. In May and June, *Idflieg* reported that comparative flight trials had demonstrated that the CL.IV was the best aircraft in the CL-category. The CL.IV arrived at the front in June 1918 and was reported as "unsurpassed by any CL-aircraft at the front."

Pilots did report slight pitch instability due to the shorter fuselage. This made the aircraft tiring to fly and less suitable as a gun platform. Halberstadt engineers solved the problem by adding a frame to the tail section, increasing the fuselage length by about 0.4 m (15.7 inches). This longer fuselage was required for the Roland-built aircraft. However, *Idflieg* delayed making the change in Halberstadt-built aircraft pending evaluation of front-line reports. It is not known if later Halberstadt-built CL.IV aircraft had the extended fuselage.

An early-production Roland-built CL.IV was static load tested between July 22 and August 14, 1918. Unfortunately, Roland engineers had 'improved' the wing structure and weakened it. The wings had to be built essentially on the original Halberstadt design. Conditional approval was granted for operational service pending a second load test. These were performed between 11–25 September. Minor structural problems were discovered requiring more strengthening. All Roland-built aircraft were delivered with the longer fuselages. All these changes made the Roland-built aircraft 20 kg heavier than Halberstadt-built aircraft.

Above: Halberstadt lineup of *Schlachtstaffel* 21. The commanding officer's CL.IV is third from right and is the aircraft depicted in the colors of the restored CL.IV in the National Museum of the USAF. The other CL.II aircraft have their black and white colors reversed; their black stripes are wider than their white stripes, giving them a darker appearance.

Above: Halberstadt CL.IV prototype on the factory airfield in February 1918. Note the old-style national insignia. The wooden mockup of a Parabellum LMG 14 shows its wide arc of fire. (Peter M. Grosz collection/STDB)

Halberstadt CL.IV Production Orders

Date	Quantity	Serials	Notes
May 1918	150	4600–4749/18	
June 1918	200	5770–5969/18	
July 1918	150	8050–8199/18	Roland
August 1918	200	6439–6589/18	Only aircraft in the 6500–6599/18 range have been seen
August 1918	200	9400–9599/18	
November 1918	300	Not known	Roland, cancelled
November 1918	100	Not known	cancelled
3 prototypes normally ordered, one for flight tests, one for static tests, and one spare.			

Above & Below: Top: Early production Halberstadt CL.IV on the factory airfield on May 1918. The lower wings of the CL.II and CL.IV were very similar but not interchangeable. (Peter M. Grosz collection/STDB)

Halberstadt CL.IV
***Schlasta* 9, tactical '6'**

Halberstadt CL.IV flown by *Lt.* Ebel of *Schlasta* 21

Halberstadt CL.IV '2' *Brünhilde*
***Schlasta* 27**

This Page & Facing Page: Halberstadt CL.IV(Rol) 8171/18 I of the first production series from Roland. The camouflage of this aircraft was that normally applied by the Roland factory, not the same as the Halberstadt factory camouflage. The name and serial number were applied with light blue paint. Most parts were stencilled with the designation *Halb CL.IV(Rol)*. (Peter M. Grosz collection/STDB)

Halb.C.L.IV(Rol.) 8171

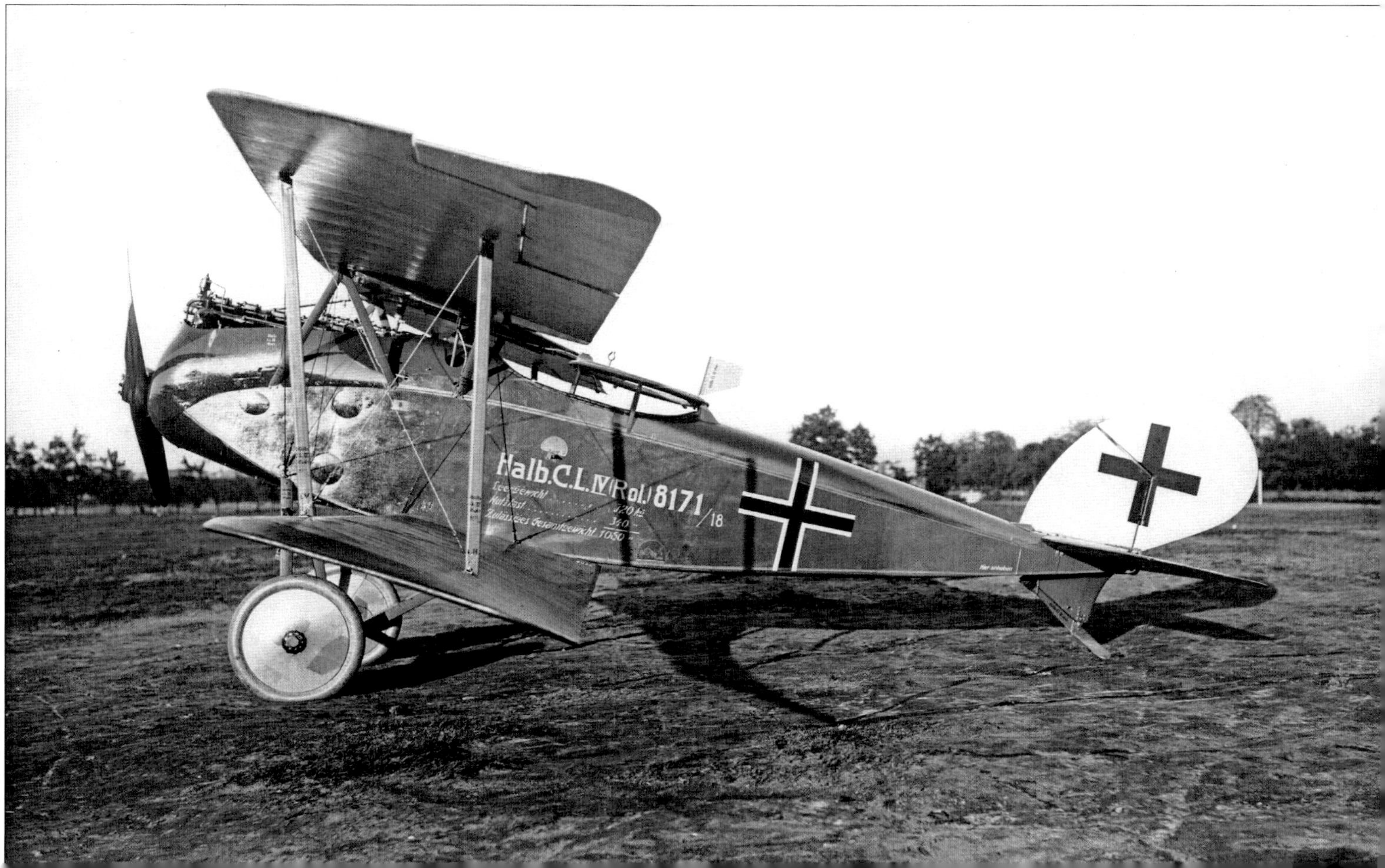
Halb.C.L.IV(Rol.) 8171/18

Above: Halberstadt CL.IV second from left from *Schlasta* 33 and Hannovers in the background of (probably) *Schlasta* 20 on August 10, 1918. Aircraft at left is a Halberstadt CL.II with another behind the CL.IV. (Peter M. Grosz collection/STDB)

Below: Halberstadt CL.IIV(Rol) 9440/18 from a manual photo showing how the seat pack version of the parachute was attached to the observer. It could also be worn as a back pack. The static line (*Zugleine*) was tied to the gun ring, (Peter M. Grosz collection/STDB)

Above: Halberstadt CL.IV '2' of an unknown unit with hammer unit insignia. On the original photo the parachute installation is seen, two metal cables as static lines were used for the crew. (Peter M. Grosz collection/STDB)

Below: This aircraft with axe insignia has a propeller-driven electrical generator. (Peter M. Grosz collection/STDB)

Above: Halberstadt CL.IV 4636/18 *Medi*, Wnr 1433, with crew. An identification streamer is attached to the lower left wingtip.

Above: Halberstadt CL.IV of the *Fliegerabteilung* of the *III Marine Brigade (Löwenfeld)* photographed at Dorsten. The written text on the lower photo refers to "Kapputsch"; no doubt this refers to the "Kapp-Putsch", which dates the photos to mid-March 1920. (Peter M. Grosz collection/STDB)

Halberstadt CL.IV
Schlasta 26 tactical '6'
Note: Thin '6' shown; thick numerals also used.

Halberstadt CL.IV(Rol) 9455/18
III Marine Brigade, tactical '2'

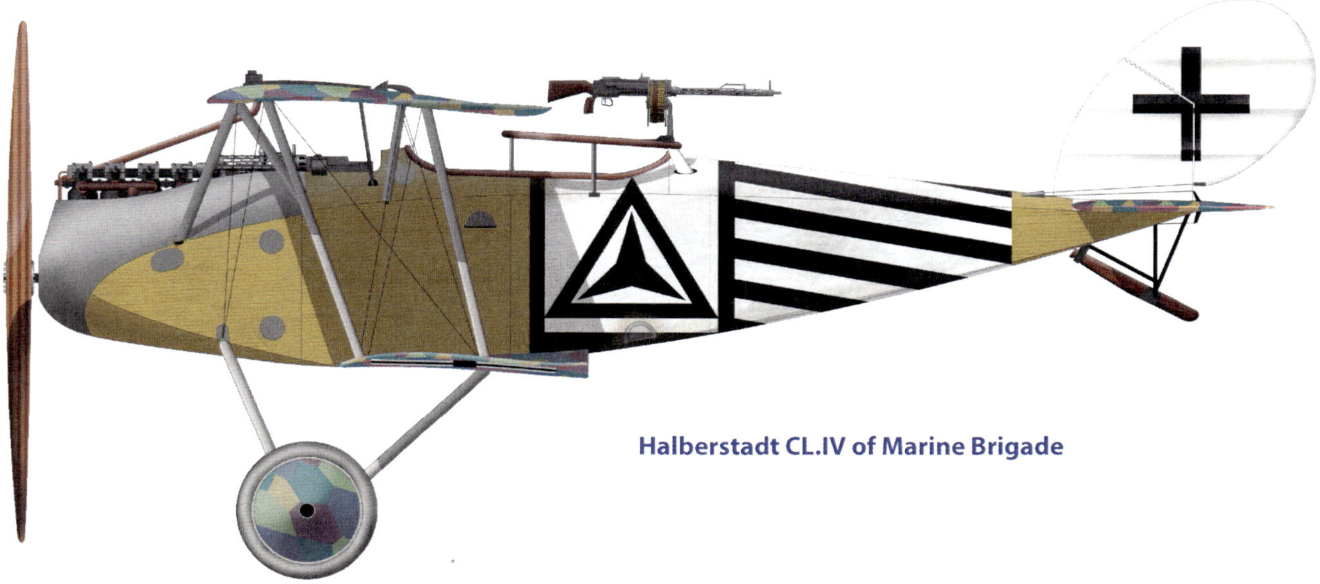

Halberstadt CL.IV of Marine Brigade

This Page & Facing Page: Halberstadt CL.IV restored in the colors of *Lt.* Ebele, commander of *Schlachstaffel* 21, on display at the National Museum of the USAF at Wright-Patterson AFB.

This Page & Facing Page: Halberstadt CL.IV(Rol) 8103/18 in the National Air & Space Museum. This aircraft was finished according to Roland factory practice; the serial and data block were applied in bright blue paint.

Halberstadt CLS.I

Above: A Halberstadt CLS.I prototype. Engine was an 180 hp Mercedes D.IIIaü. (Peter M. Grosz collection/STDB)

Both Hannover and Halberstadt produced successful light C-types as two-seat fighters. The CLS.I was Halberstadt's planned successor to the earlier CL.II and CL.IV. In fact, the CLS.I looked like a refined CL.IV with slightly different proportions. The engine was a 180 hp Mercedes D.IIIaü, a high-altitude version of the reliable engine that powered so many fighters and the previous CL.II and CL.IV.

Aircrew of the CL-types carried extra weapons and ammunition to carry out their missions, including hand grenades, small bombs, and extra drums of machine-gun ammunition. This overloaded the CL-types beyond their designed useful load limit and caused structural damage to the airframes. When *Kogenluft* addressed this issue on a conference on September 6, 1918, both *Idflieg* and Halberstadt were found blameless.

Kogenluft decided that the useful load rating of CL-class *Schlachtflugzeuge* (infantry-support aircraft) would be raised from 346 kg to 420 kg and the CL-class design specification would be revised to meet the strength requirements (maximum G limits) of fighter aircraft.

Furthermore, CL-type design emphasis was to be focused on speed, maneuverability, and diving ability at the expense of a fast rate of climb. This reflected the fact that the role of the CL-types had expanded and their use as *Schlachtflugzeuge* was now a critical role.

Halberstadt CLS.I Specifications

Engine:	180 hp Mercedes D.IIIaü	
Wing:	Span	9.7 m
	Chord, upper	1.600 m
	Chord, lower	1.400 m
General:	Length	6.95 m
	Empty Weight	670 kg
	Payload	520 kg
	Loaded Weight	1190 kg
Maximum Speed:		185 km/h
Climb:	4500m	36 min
Fuel Capacity:	Main tank	60 Liters
	Secondary tank	48 Liters
Armament:		3 mgs

Halberstadt CL.IV aircraft revised to withstand the higher load rating began to come off the production line in September 1918. At the same time, Karl Theis, the designer, was working on the CLS.I, an aircraft designed from the beginning for the stress limits of the *Schlachtflugzeuge* role.

CLS.I armament was two synchronized Spandau guns for the pilot and one Parabellum for the gunner. A 3.5 cm revolver cannon was said to be planned for the CLS.II and CLS.X but this was a plan only.

Despite having a shorter wing-span than the CL,IV, the

Above & Below: A Halberstadt CLS.I prototype. The aircraft below has no tires. (Peter M. Grosz collection/STDB)

CLS.I had a heavier loaded weight reflecting the more robust structure and heavier weight of weapons and ammunition.

The CLS.I was type-tested on 2 October 1918, commendably soon after the changed requirements. It was intended to replace the CL.II and CL.IV but the Armistice intervened.

Serial numbers 1797/18 and 1799/18 are known, and their were probably others. One of these was work number 2105.

Above & Below: A Halberstadt CLS.I prototype. (Peter M. Grosz collection/STDB)

Halberstadt CLS.I Prototype

Above: A Halberstadt CLS.I prototype. (Peter M. Grosz collection/STDB)

Below: Halberstadt CLS.I fuselage in the Deutsches Technikmuseum in Berlin. (Author's photo)

Hannover CL.II

Above & Below: Hannover CL.II C4501/17 was one of three production prototypes photographed during its *Typenprüfen* at Adlershof in July 1917. Three prototypes were usually ordered, one for load testing, one for flight testing, and a spare.

Idflieg issued requirements for new categories of combat aircraft in September 1916 as a result of learning the lessons of the Battles of the Somme and Verdun. Those new categories included a light C-type for ground attack and escorting conventional C-types. This new category, called CL for light C-type, was basically a two-seat fighter.

Targeting the new CL-category for their first original design, Hawa and Dorner had essentially a free hand with the exception that *Idflieg* specified use of the 180 hp Argus As.III engine. The Argus was reliable but its performance fell off rapidly at higher altitudes, so fighter producers preferred – and usually received – the superior 160 hp Mercedes D.III. However, the new CL-type was expected to operate at low and medium altitudes, not exceeding 3,000 m, for which the Argus was considered suitable. With the Argus being built by six different companies and in limited demand – most two-seaters being built were using the 200 hp Benz and fighters were using the 160 hp Mercedes – availability of the Argus for CL-types was assured.

Dorner's first design for Hawa was originally designated the Hannover C.II; this was later changed to Hannover CL.II to indicate it belonged to the CL-category. The Aviatik C.I(Han) had originally been known as the Hannover C.I until *Idflieg* rationalized type designations starting in August 1916.

The CL.II was of conventional wood, wire, and fabric construction with the fuselage covered with thin plywood for strength and streamlining. Dorner exercised great attention

Above: Hannover CL.II C4501/17 photographed during its *Typenprüfen* at Adlershof in July 1917. The photograph shows the attention to streamlining for this single-bay biplane.

to detail and the resulting airframe was strong, light weight, and streamlined for a biplane. A fixed, synchronized gun was provided for the pilot and the gunner had a flexible machine gun. The fuselage was deep, providing good protection for the crew and giving the gunner a wide field of fire. Unusually, the tail was of biplane configuration to minimize the obstruction to the gunner's field of fire to the rear.

The CL.II passed its *Typenprüfung* (type test) on 21 July 1917 and demonstrated excellent flight qualities and performance. This was followed in August 1917 by the first order of 200 aircraft from *Idflieg*.

The CL.II was instantly successful at the Front. Not only

Above: Hannover CL.II 9303/17, photographed at Les Baragues, shows the shape of the wingtips of early production CL.II machines. The gravity tank is to port on the centre section with the radiator just starboard of the center-line.

Hannover Aircraft Orders & Production, 1915–1919			
Type	**Known Serials**	**Quantity**	**Comments**
Aviatik C.I(Han)	1951–1974/15 600–649/16 2100–2123/16 2764–2787/16 3367–3390/16	24 50 24 24 24	1951/15 type test February 1916 2114/16 type test August 1916 Total 146 built
Halberstadt D.II(Han)	800–829/16	30	800/16 type test October 1916
Rumpler C.Ia(Han)	4600–4799/16 900–974/17 5475–5574/17	200 75 100	4600/16 type test January 1917 Ordered October 1916 Ordered October 1916
Hannover CL.II	4500–4502/17 9200–9399/17 13075–13317/17	3 200 243	Ordered in April 1917 Ordered in July 1917 Total of 446 built
Hannover CL.III	16000–16079/17 Built after Nov. 1918	80 100	Type test February 1918 Total of 180 built; 80 built during the war
Hannover CL.IIIa	13318–13374/17 16080–16199/17 2600–2799/18 3800–3899/18 6950–7049/18 9600–9649/18	57 200 200 100 100 50	57 Han. CL.II aircraft were burned in a factory fire that did not get works numbers or military serials. Apparently they were replaced by CL.IIIa aircraft using the missing serials in the C.13xxx/17 batch as shown here. Total of 707 built
Hannover C.IV	—	5(1)	Type test March 1918
Hannover CL.V	9650–9699/18 Built after Nov. 1918	50(2) 62	9651/18 type test October 1918 Total of 108 built
Hannover Type 10	NA	1	Built in 1919

Notes: 1. Often quoted as only 1 C.IV built. 2. 46 of these 50 were built. Highest confirmed serial 9680/18.
3. In 1918 200 Han. CL.II(Rol) serials 500–699/18 were built by Roland.
4. In total Hannover built 1,899 aircraft. The table adds up to 1956 aircraft without the one Type 10; the difference between these numbers is the 57 CL.II burned in the factory fire.

Hannover CL.II 13189/17
FA(A) 287b

Hannover Aircraft Specifications				
Type	**CL.II**	**CL.III**	**CL.IIIa**	**CL.V**
Engine	180 hp Argus As.III	160 hp Mercedes D.III	180 hp Argus As.III	185 hp BMW.IIIa
Span, m	11.95	11.70	11.70	10.49
Length, m	7.80	7.58	7.58	7.00
Height, m	2.75	2.80	2.80	2.84
Wing Area	33.8 m^2	32.7 m^2	32.7 m^2	28.5 m^2
Empty Wt., kg	750	741	750	720
Useful Load, kg	360	340	360	360
Gross Wt, kg	1,110	1,081	1,110	1,080
Max Speed, km/h	165	160	165	175
Time to 1,000 m	6.1 minutes	5.3 minutes	—	3.3 minutes
Time to 2,000 m	13.8 minutes	11.1 minutes	—	7.0 minutes
Time to 3,000 m	23.9 minutes	19 minutes	—	12.0 minutes
All the above types carried a fixed, synchronized gun for the pilot and a flexible gun for the observer.				

Hannover CL.II 13081/17
FA(A) 226, Tactical 'K'
Vzfw. Willy Engler and *Lt.* Alfred Kuemann

was it strong and easy to fly, it had excellent maneuverability. It had the ability to turn sharply without losing altitude and remarkably could turn inside the Sopwith Camel or SE.5a fighters, which side-slipped in steep turns whereas the CL.II did not. Of course, the CL.II had the advantage of a rear gunner in addition to the pilot's fixed gun, which, combined with its strength and maneuverability, made it very effective in air-to-air combat. Its crews discovered they could accept and break off combat with Allied fighters at will. In air combat the CL.II was considered more effective than either the DFW C.V or Rumpler C.IV, especially 'the ease with which they could be spun to avoid enemy fighters and Flak.'

The Hannover CL.II could carry cameras and wireless like the standard C-types, but were not usually employed for these tasks. Typically the CL.II was used to escort regular C-type aircraft or combined into *Schutzstaffeln*, or *Schustas*, which were protection flights to guard other two-seaters. These were later converted to *Schlachtstafflen*, which were primarily ground-attack units.

Above: Hannover CL.II 'diamond 3' in flight.

Above: Hannover CL.II 13274/17 aircraft "4" of *Schlasta* 25 at *Armee Flug Park* III at Ghent, April/May 1918. Note the unusual white tailplane adopted by this unit.

Above: Hannover CL.II 9237/17 ffom the first production batch. Note the Mercedes-type three-pointed star insignia of the fuselage top and sides over the dark stripe. The unit is unknown.

Above: Hannover CL.II 9390/17 ffom the first production batch in the left foreground lined up with the rest of its unit.

Hannover CL.II C.9276/17
Schusta 27b Tactical '5'
Left Side

Hannover CL.II C.9276/17
Schusta 27b Tactical '5'
Right Side

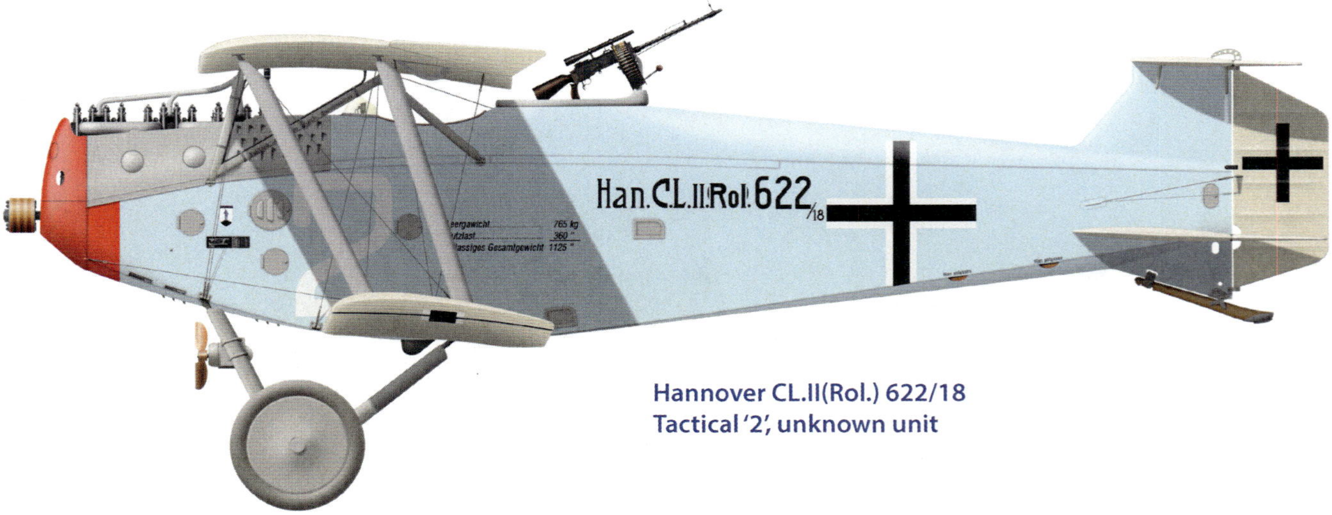

Hannover CL.II(Rol.) 622/18
Tactical '2', unknown unit

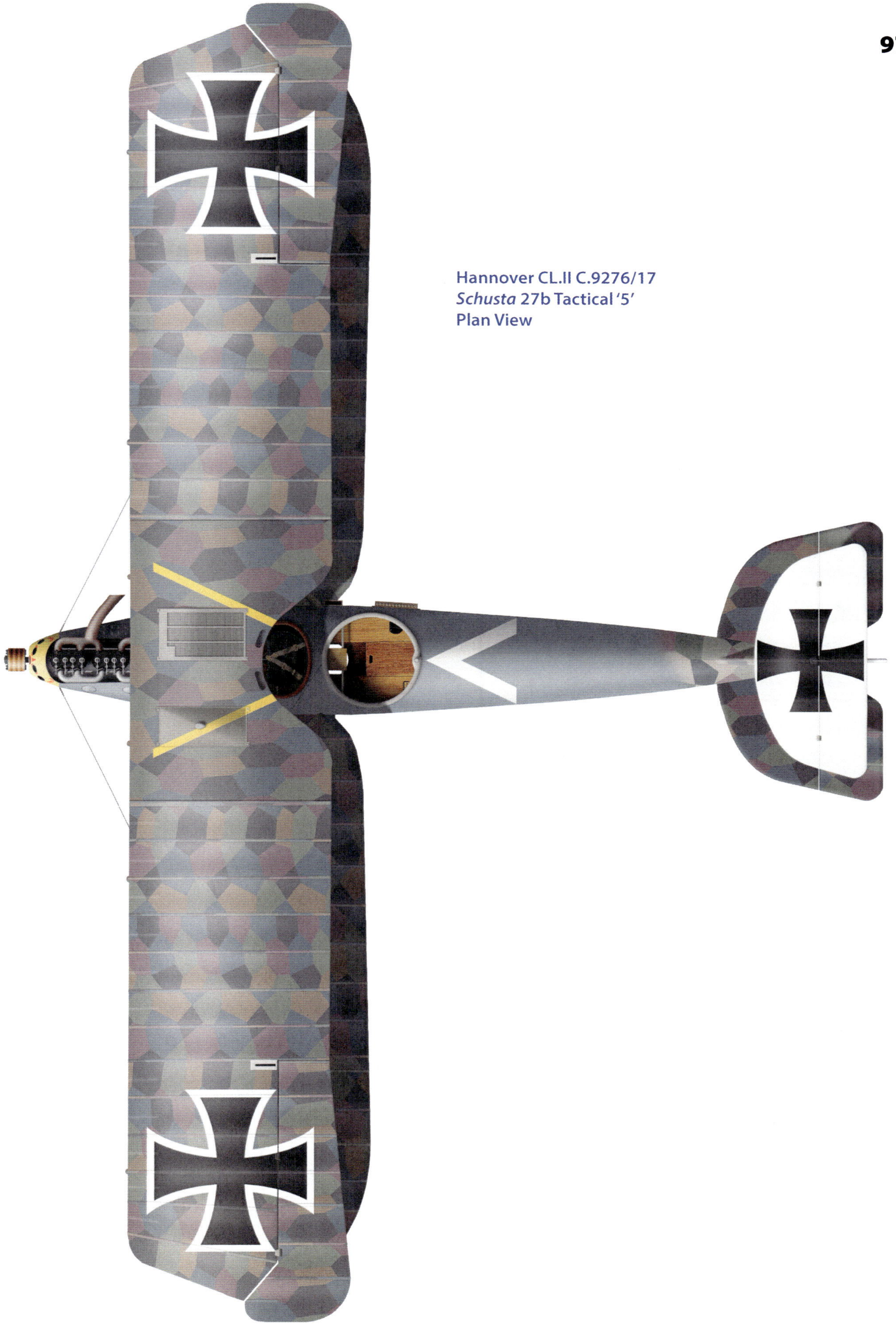

Hannover CL.II C.9276/17
Schusta 27b Tactical '5'
Plan View

Above & Below: Hannover CL.II 9276/17 tactical number 5 of *Schusta* 27b. Note the addition of a cross to the top of the upper tailplane.

Hannover CL.II 13080/17
Unit unknown

Hannover CL.II 9232/17
Schlasta 12, Tactical '5'
Left side

Hannover CL.II 9232/17
Schlasta 12, Tactical '5'
Right side

Hannover CL.III & CL.IIIa

Above: This unmarked Hannover CL.III may have been the prototype. The CL.III was powered by a 160 hp Mercedes D.III engine. By 1918 the new Mercedes engines being built were 170 hp Mercedes D.IIIa engines but usually this distinction is ignored. Only 80 CL.III could be delivered during the war due to the shortage of Mercedes engines needed for fighters.

In early 1918 Hawa delivered an improved aircraft, the CL.III, powered by the 160 hp Mercedes. The CL.III was a refinement of the CL.II and the two types were very difficult to distinguish from one another.

Compared to the earlier CL.II, the airframe of the CL.III was further strengthened and structural weight was reduced, partly by reducing overall dimensions somewhat.

The CL.III passed its *Typenprüfung* (type test) on 23 February 1918 and *Idflieg* ordered the CL.III in March. Due to the shortage of Mercedes engines which were desperately needed for fighters, only 80 CL.III aircraft were built during the war.

As a result, to continue delivering aircraft, Hannover had to revert to using the Argus engine, the resulting aircraft being designated the CL.IIIa.

The CL.III and CL.IIIa arrived at the Front in March–April 1918. Like the earlier CL.II, they were immediately successful and well-liked by their crews. The aircraft were fully aerobatic, very maneuverable, and featured good take-off and landing qualities, which was important to front-line aircrews.

Shortly after the aircraft arrived at the front a few wing failures were experienced under the rough field conditions and *Idflieg* ordered strengthened wing fittings on the aircraft in production. Furthermore, the load requirements for CL-class aircraft were later upgraded to that of fighter aircraft, which was 5 Gs.

The improved CL.IIIa no longer experienced wing flutter or vibration.

Above & Below: Hannover CL.IIIa 3879/18 from the third CL.IIIa production batch shows the characteristic valve springs of the Argus As.III engine. The balanced ailerons of the CL.III & CL.IIIa reduced the pilot's control forces and improved maneuverability compared to the CL.II with plain ailerons.

Hannover CL.IIIa, Unit Hack of *Jasta* 17
Flown by *Jasta* C.O. *Lt.* Günther Schuster and others,
June–July 1918

Hannover CL.III Specifications

Source	PMG Data 16000/17	DSA Data
Type	**CL.III**	**CL.III**
Dimensions in m		
Span upper	11.70	11.700
Span lower	11.02	11.020
Chord upper	1.80	1.800
Chord lower	1.30	1.300
Length	7.80	7.580
Height		2.800
Stagger	0.77	0.800
Gap	1.51	
Wheels		760x100
Airscrew		
Type		Zentral
Diameter in m		2.750
Area in m²		
Wings	32.7	
Aileron		
Horizontal Stabiliser Upper/lower		
Weights in kg		
Empty	741	741
Loaded	1,081	1,081
Crew		
Armament		
Instruments		
Fuel & Oil		
Motor		
Exhaust		
Instruments		
Radiator		
Water		
Airscrew		
Fuel tanks		
Oil tank		
Motor accessories		

Notes: * Empty weights for 16000/17 = 736 kg
16018/17 = 748 kg
16001/17 = 753 kg

Hannover CL.III Specifications (cont.)

Source	PMG Data 16000/17	DSA Data
Speed in kph		
Max		165
Landing		
Climb in min		
to 1,000 m	5.3	
to 2,000 m	11.1	
to 3,000 m	19.0	
to 4,000 m	27.6	
to 5,000 m		
Engine	180-hp Mercedes	180-hp Mercedes D.IIIaü

Hannover CL.IIIa Specifications

Source	PMG Data	DSA Data
Dimensions in m		
Dimensions in m		
Span upper	11.70	11.700
Span lower	11.02	11.020
Chord upper	1.80	1.800
Chord lower	1.30	1.300
Length	7.80	7.580
Height		2.800
Stagger	0.77	0.800
Gap	1.51	
Wheels		760x100
Airscrew		
Type		Reschke
Diameter in m		2.850
Weights in kg		
Empty	760	750
Loaded	1,120	1,110
Speed in kph		
Max		165
Climb in min		
to 1,000 m	3.2	
to 2,000 m	7.1	
to 3,000 m	11.9	
to 4,000 m	18.3	
to 5,000 m	28.0	
Engine	Argus As.III	Argus As.III

Identification Features of Hannover CL.II & CL.III	
Aircraft	**Notes on Visible Features**
Prototype Argus As.III	Unbalanced ailerons. Aileron crank at 1/3 span aileron. Splayed-out two outer ribs on top wing. Four ribs outside of interplane struts on upper wing, five on lower wing. Engine cowl changed on later versions, twin upper tail braces. (See *Flugsport* 1919, P.417 for side view).
CL.II Argus As.III Three aircraft for *Typenpüfung* CL 4500–4502/17.	Aileron does not project beyond wing. Tail square on prototype. Aileron control wires outside of strut. Aileron crank at ½ span aileron. Twin upper tail braces. Balanced elevator. Four ribs outside of interplane struts on upper wing, five on lower wing. Splayed out outer upper wing tip ribs. Dihedral equal top and bottom wings. Modified engine cowl compared with prototype. Cooling louvres below centre section struts.
CL.II early production Argus As.III CL 9200 – 9399/17	Balanced aileron does not project past wing. Aileron crank at aileron end. Rectangular fuselage vents on port side. Tail rounded. Aileron control wire inside of strut. Twin upper tail braces.
CL.II 9268/17 & up	Cooling louvre on access port panels. Spreader bar for interplane wires.
CL.II 13075–13374/17	Balanced aileron projected past wing. Braced tail on almost all of this series. Spreader bar for interplane wires. *CL 13318–13374 burned in fire. Made up by CL.IIIa with same serials.*
CL.III Mercedes 16000 – 16079/17	Enlarged tail brace. No cooling louvres on port fuselage side.

Above: Hannovers were tough aircraft and so were many of their crews. On one occasion pilot *Vzfw*. Johann Baur and observer, *Lt*. Goerg van Hengl of *FAA* 295 were flying their CL.IIIa 2622/28 on an artillery spotting mission. Arriving over their briefed location, they were annoyed to find the airspace already occupied by a flight of six French Spad 13 fighters. Unable to conduct their mission with the Spads in the same location, they immediately attacked the fighters. After claiming two Spads and driving off the other four, they conducted their mission of spotting for the artillery. Postwar Johann Baur became Hitler's personal pilot, an experience he wrote about in his autobiography *Hitler by My Side*.

Above: Hannover CL.III 16059/18 from the only CL.III production batch produced during the war. The Mercedes engine that powered the CL.III was in great demand for single-seat fighters and the CL.IIIa reverted to the Argus engine. It bears the thick interim style German national markings on the fuselage with the early curved crosses on the wings.

Above: Hannover CL.IIIa.

Above: Hannover CL.IIIa 3885/18 of *Schlasta* 3 landed in US lines on 3 October 1918 and was moved by German PoWs under trees to hide it from enemy artillery, which would have shelled it if it was seen.

Above: Another view of Hannover CL.IIIa 3892/18 after capture. The broad arrow on the fuselage was the unit insignia of *Schlasta* 20. The "Trophy" legend painted on this machine has led to it being misidentified as the aircraft brought down by Rickenbacker and Chambers.

Above: Hannover CL.IIIa with an interesting marking waits for its next mission.

Right: Hannover CL.II(Rol) 622/18 bears a white individual numeral "2" on the forward fuselage and a dark painted nose. Note the auxiliary struts to the tailplane.

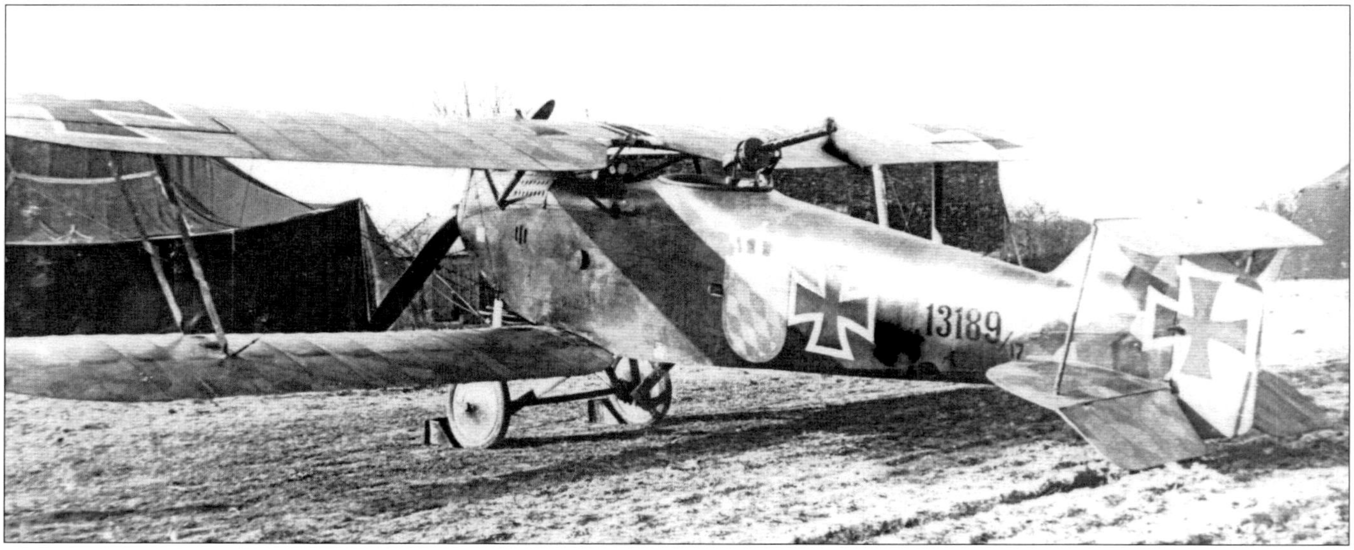

Above: Hannover CL.II C.13189/17 of *Fl.Abt(A)* 287b displays the Bavarian *Rauten* symbol with crown on the fuselage.

Above: Hannover CL.II '5' of *Schlasta* 12. The white-black-white stripe is the unit marking. Note the rack holding grenades and the curved rod to prevent the gunner from shooting through the propeller arc. The band continued around the fuselage to the bottom of the starboard wing,

Above: Hannover CL.II 13080/17 with a 'reddy kilowatt' marking with crewmen of its unknown unit.

Hannover CL.IIIa 27XX/18
FA(A) 253
Uffz. Braunger & *Lt.d.R.* Willy Hentschel
1 August 1918

Hannover CL.IIIa 2714/18
FA(A) 253
Uffz. Peter Johanner & *Lt.d.R.* Hanns-Gerd Rabe
Autumn 1918

Hannover CL.IIIa 2622/18 of *FAA* 295
Flown by *Lt*. Georg von Hengl and *Vzfw*. Johann Baur,
Summer–Autumn 1918

Above: *Leutnant der Reserve* Hanns-Gerd Rabe (right) and fellow observer *Ltn.d.Reserve* Willy Hentschel seen in front of the Hannover CL.IIIa that Hentschel and his pilot, *Unteroffizier* Braunger, used in shooting down a Sopwith Camel west of Givenchy on 1 August 1918. The victory confirmation was published in a *Kofl 6. Armee* report and acknowledged in the *Nachrichtenblatt der Luftstreitkräfte*. (via Peter Kilduff)

Above: *Flieger-Abteilung (A)* 253 observer *Leutnant der Reserve* Hanns-Gerd Rabe (left) and his pilot, *Unteroffizier* Peter Johannes (middle), flew low-level reconnaissance missions in Hannover CL.IIIa 2714/18, seen here. The "bird" insignia on the fuselage is the symbol of the *Wandervogel* [Bird of Passage] youth group of which Rabe was once a member; the symbol also alludes to the German word *Rabe* [Raven]. Their aircraft mechanic, *Uffz.* Noll, stands at right. (via Peter Kilduff)

Hannover CL.V

Above: The CL.V was first tested at Aldershoft with the biplane tail. This prototype was powered by the 170-hp Mercedes D.IIIa engine.

The next Hawa design, its final design of the war, was the CL.V, powered by the 170 hp Mercedes D.IIIa. The uncovered CL.V airframe was inspected in June 1918, and it was hoped the first flight would occur in July. Although intended to replace the CL.IIIa in production, the Hawa CL.V was based on a scaled down, refined version of the C.IV airframe. Its robust cabane struts were single, streamlined struts on each side that were integral with the aircraft's fuselage. The interplane struts were I-struts designed for good field of view and reduced drag. Fabric-covered duralumin control surfaces were used. A nose radiator was employed and the aircraft retained the biplane tail that characterized the earlier Hawa CL-types and which had been adopted to improve the gunner's field of fire.

The CL.V was load-tested in July with the biplane tail. However, at Hawa's request, a CL.V prototype with conventional tail was substituted. The tests revealed insufficient strength of the elevator, rudder, undercarriage, and center-section. Nevertheless, 20 CL.Vs powered by the 170 hp Mercedes were ordered for front-line evaluation with the requirement that the first production aircraft be tested again to verify it met the strength requirements. Upon retest the *Typenprüfung* was successfully passed on 7 October 1918.

However, in September *Idflieg* had decided that the Hawa CL.V order would not be increased. The reason was that *Idflieg* had issued two new class specifications for *Jadgzweisitzer* (two-seat fighter) and a light *Schlachtfluzgzeug* (ground-attack aircraft) to address requests from the commander of the German Air Service.

The *Jadgzweisitzer* concept had evolved from combat experience and front-line demands. The Hawa CL.V was recognized as having the best performance and flight characteristics for its power. When powered by the 185 hp BMW.IIIa engine it was as fast as the BMW-powered Fokker D.VII, the best German fighter. Furthermore, the CL.V had excellent maneuverability, fully as good as the Albatros D.III in mock combat. It was realized that was not yet good enough but improvement was expected from tuning the airframe. As a result of the tests, it was recommended that 20 CL.Vs powered by BMW.IIIa engines be ordered to the *Jadgzweisitzer* requirement.

Essentially, the *Jadgzweisitzer* concept was simply an evolution of the original escort fighter concept of late 1916 that resulted in the CL-category. With three guns, two synchronized for the pilot plus a flexible gun for the observer, plus fighter-like performance and airframe strength, the

Above: The first version of the Hannover CL.V with biplane tail. This front quarter view shows the nose radiator and streamlined cabane struts to advantage.

Jadgzweisitzer was expected to penetrate enemy airspace and meet enemy fighters on basically even terms. If the war had continued the CL-category of aircraft would have been built to *Jadgzweisitzer* specifications rather than establishing a separate category. After the first batch of 20, all subsequent CL.V aircraft were built to *Jadgzweisitzer* standards and load-tested to single-seat fighter specifications.

The light *Schlachtflugzeug* class (CLS) was established in September 1918 for aircraft designed for ground-support. The CLS specification required a higher useful load – meaning more ammunition and 50 kg of anti-personnel bombs. For this class the requirements were higher speed, maneuverability, and diving capability rather than climb rate. Like the new CL-aircraft the CLS-types were to be built and tested to the single-seat load requirements. The Hawa CL.V may have fulfilled the CLS role also.

A total of 108 Hannover CL.Vs were built, 46 before the Armistice and 62 after, and many of these were obtained by the air services of the new nations created as a result of the war's end.

However, the CL.V saga does not end there. Norway chose the Hannover CL.V for license production at the Kjeller Flyvermaskinfabrikk in Hal-den. Under the designation F.F. 7 *Hauk* ("Hawk"), 14 aircraft were built in 1923–1924. These aircraft were powered by the 185 hp BMW.IIIa engine and were used as reconnaissance fighters and advanced trainers. The *Hauk* lasted in Norwegian service until 1929.

Right: Norwegian-built *Hauk* serial 225 has the Norwegian markings around the fuselage rather than on the rudder, as on the *Hauk* in the background. This marking variation is seen on a number of Army and Navy aircraft. The reason for this is unknown. Postwar, Norway built 14 Hannover CL.V aircraft under license; these aircraft, called *Hauk* (Hawk) by Norway, were powered by BMA.IIIa engines. (Norsk Teknisk Museum)

Above: The second version of the Hannover CL.V design replaced the biplane tail of the earlier Hawa types with a more conventional tail. This was done at the request of Hawa.

Above: The final Hannover CL.V design used a conventional, monoplane tail. This image shows the refined, streamlined lines of the CL.V. The pilot's fixed guns are clearly visible.

Above & Below: This is the final version of the CL.V design. The nose radiator had been perfected and was used to good effect on many German aircraft in the latter days of the war.

Above & Below: Airbrushed photo of the second version of the Hannover CL.V design showing nose details. Closeup below.

Above: The second version of the Hannover CL.V design with monoplane tail under construction in the factory.

Above: The final version of the Hannover CL.V with monoplane tail with a Hannover CL.IIIa in the background. This shows the CL.V was more streamlined that the earlier Hannover designs.

Hannover CL.IV Specifications

Source	1.	2.	3.	4.		5.	6. F 6	7. F.F.5 Hauk
Dimensions in m								
Span upper	10.56	10.5	10.490	10.49	10.49	10.5	10.490	10.50
Span lower	10.46							
Chord upper	1.600						1.600	
Chord lower	1.350						1.315	
Length		7.0	7.000	7.00	7.00	7.1	7.000	6.98
Height		2.85		2.84	2.84	2.9	2.840	2.80
Track							1.800	
Airscrew Type								Argus
Airscrew dia			2.900				2.900	2.80
Areas in m^2								
Upper wing								16.072
Lower wing								12.712
Total wing		28.5		28.5	28.5	28.5		
Stabilizer								1.550
Weights in kg								
Empty		720		720	720	720		696
Loaded		1,080		1,080	1,080	1,080	1,050	1,080
Capacities in litres								
Fuel								
Oil								
Performance								
Speed in kph								
Max	180		165	165	175		180	170
Cruising								100
Landing	90							70
Climb to								
1,000 m				3.3	3.3			
2,000 m				3.8	3.7	3.7		
3,000 m				5.2	5	8.7		
4,000						15.7		
5,000						25.7		
Radius in km								
Engine				160-hp Mercedes D.IIIa	185-hp BMW IIIa	160-hp Mercedes or 180-hp BMW		185-hp BMW IIIa

Source: 1. *Idflieg* drawing.
2. *Hawa-Nachrichten*, Vol.3 No.3 Dec 1920.
3. Hannover drawings.
4. P.M. Grosz data.
5. *Die Deutschen Lüftstreitkräfte im Weltkriege.*
6. Hawa pamphlet.
7. Hafsten, B. „Hauk“. *Norsk Flyhistorisk Tidsskrift*, No.2 1993

Hannover Production			
Type	**Serials**	**Quantity under Contract**	**Notes**
Aviatik C.I (Han)	1951 – 1974/15	24	C.1951/15 type tested February 1916.
	600 – 649/16	50	
	2100 – 2123/16	24	C.2114/16 type tested August 1916.
	2764 – 2787/16	24	
	3367 – 3390/16	24	Total 146 built.
Halb D.II (Han)	800 – 829/16	30	D.800/16 type tested October 1916.
Ru C.Ia (Han)	4600 – 4799/16	200	C.4600/16 type tested January 1917.
	900 – 974/17	75	Ordered October 1917.
	5475 – 5574/17	100	Ordered October 1917.
Han CL.II	4500 – 4502/17	3	Prototypes ordered May 1917.
	9200 – 9399/17	200	1st production batch, ordered September 1917.
	13075 – 13317/17	300	57 are believed destroyed in the factory fire of March 1918. 243 completed. The 57 were rebuilt as CL.IIIa machines (13318-13374/17).
Han CL.II (Rol)	500 - 699/18	200	Licence built trainers by LFG Roland.
Han CL.III	1600 – 16079/17	80	Ordered January 1918. 160-hp Mercedes D.III or D.IIIa. Type test February 1918.
Han CL.IIIa	13318 – 13374/17	57	The 57 CL.II aircraft destroyed are believed to have been replaced by this batch of CL.IIIa machines.
	16080 – 16199/17	120	Ordered January 1918. Last of order for 200 delivered as CL.IIIa
	2600 – 2799/18	200	Ordered April 1918.
	3800 – 3899/18	100	Ordered May 1918.
	6950 – 7049/18	100	Ordered February 1918.
C.IV	-	2	At least two built. Type tested March 1918. (P.M. Grosz gives 5 as number of C.IV biplanes.)
CL.V	9800 – 9802/18	3	Prototypes.
	9650 – 9699/18	50	Ordered September 1918. CL.9680/18 is highest known serial. 62 were built after Armistice.
Type 10	Built postwar	1	Civil triplane.
Thanks to Reinhard Zankl for the above compilation.			

Left: The final version of the Hannover CL.V design with monoplane tail in flight.

Facing Page, Top: Closeup view of the Halberstadt CL.V. cockpit area.

Facing Page, Bottom: The central portion of a one-piece Hannover CL.V wing has been cut out showing the aileron control rod cutouts and the gravity tank.

Below: Rear view of monoplane tail version of the Hannover CL.V.

Bottom: First version of the CL.V with biplane tail.

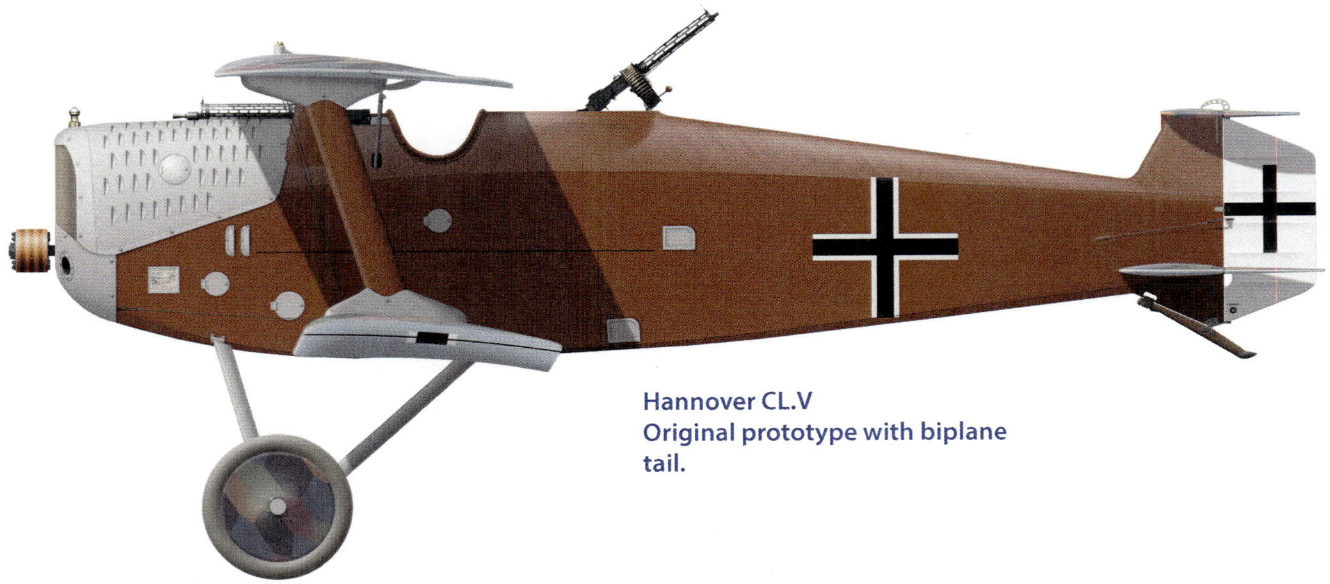

Hannover CL.V
Original prototype with biplane tail.

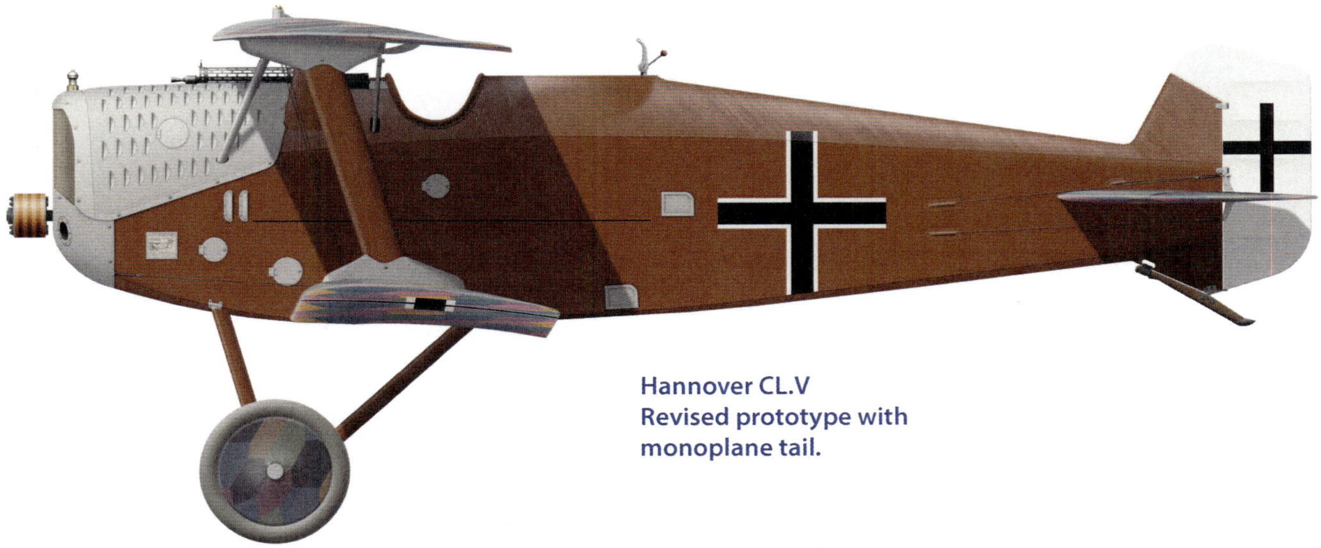

Hannover CL.V
Revised prototype with monoplane tail.

Hannover CL.V 9671/18
Postwar Polish Service

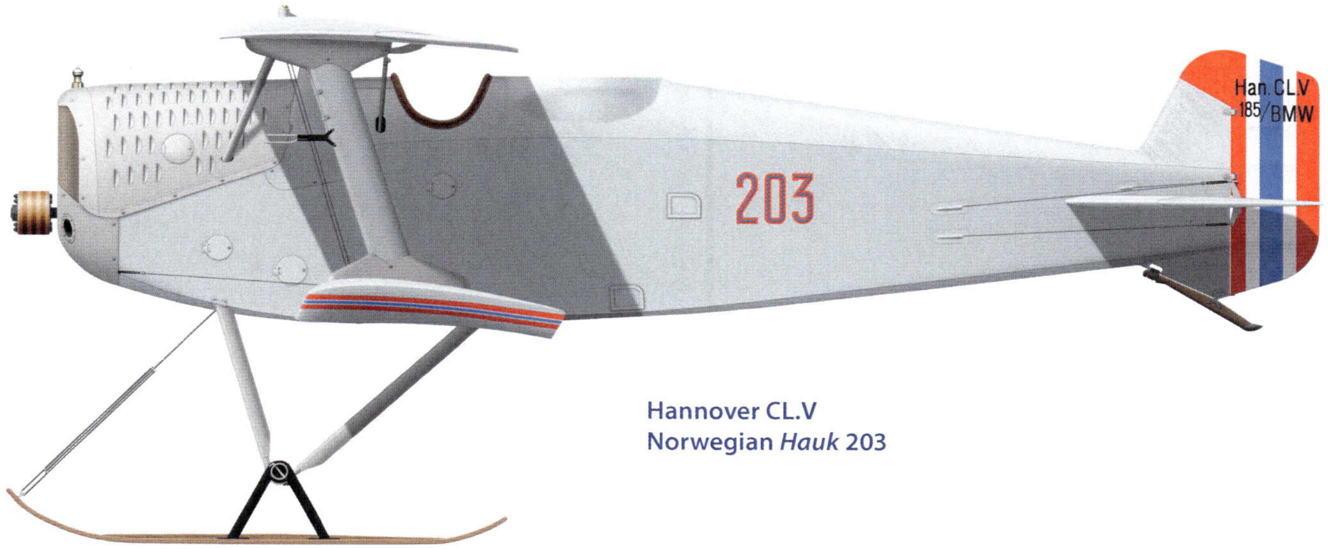

Hannover CL.V
Norwegian *Hauk* 203

Hannover CL.V
Norwegian *Hauk* 213

Hannover CL.V
Norwegian *Hauk* 225

Hannover CL.V *Hauk* in Norway

Above: Hannover CL.V *Hauk* 213 with pilot. (Norsk Teknisk Museum)

Below: Hannover CL.V *Hauk* 203 with pilot. (Norsk Teknisk Museum)

Above: Another view of 203 displaying the ski arrangement. Note the ski on the tail skid. (Norsk Teknisk Museum)

Below: Hannover CL.V *Hauk* 207 on skis with pilot aboard and engine running. (Norsk Teknisk Museum)

This Page: Hannover CL.V *Hauk* on display at the Norsk Teknisk Museum.

Junkers CL.I

Above: Junkers J 8 on 2 January 1918, the CL.I prototype. Note the balanced ailerons and the shape of the wing tips. The landing gear has two inverted V-struts from the fuselage centreline to the outer ends of the axle supports, and a single central strut. The wheel covers have "Junkers & Co Dessau" stencilled on them. The port wing root is strengthened with a walkway as the corrugated metal could not stand up to rough treatment. (AHT AL0459-220 and SDAM)

Hugo Junkers started developing all-metal airplanes early in the war. Junkers preferred the simplicity of monoplanes, and his first airplane was a monoplane. The first Junkers type that went into production was the J.I (military designation; J 4 internal company designation) sesquiplane, a configuration required by the army. The J.I was the world's first production all-metal airplane.

Junkers continued developing new types, including the J 7 monoplane single-seat fighter. The J 8 was a two-seat monoplane developed from the J 7. It had a Junkers nose radiator and large balanced ailerons. The observer was armed with a Parabellum on a gun ring and the pilot had a synchronised machine gun. Power was supplied by a Mercedes D.III engine. It was designed for close infantry support to replace the Halberstadt and Hannover CL aircraft used in these operations but was not armored.

Most references state that only one J 8 was constructed, it serving as a prototype for the production J 10, known as the Junkers CL.I when placed in prodion.

The J 8 was first flown on 10 December 1917, from Halle airfield, a year after the contract was let. That same month Fokker wrote to Professor Junkers that the two partners in *Jfa* were committed to develop a C-class machine (that is a two-seat machine with armament for the pilot and observer). Fokker stressed the advantages of the J 8 with its field of fire for the observer and its high speed, concluding that it would have many uses at the Front.

Fokker flew the J 8 on the 14th, a few days after he damaged the J 7. He was impressed with its performance. Fokker's test pilot, Schmidt, then flew tests for climb rates and speeds on 8 January 1918. During this flight fog rolled in, obscuring the landing field. Schmidt then experienced engine failure when the fuel feed to the engine failed, and he had to make an emergency landing. Fortunately, discovering a hole in the fog, he put the J 8 down in a field but the machine turned over. Schmidt was unhurt and the monoplane suffered minor damage.

Idflieg recorded in its February 1918 monthly report that the machine had been developed by *Jfa* and the aircraft, that was powered by a 160-hp Mercedes, had gone through the type acceptance tests. A final judgement on the machine's flight characteristics could not be made at that point in time but the firm had been given a test order for 20 machines in order to gather experience with this type.

The J 8 accompanied the J 7 to the First *D Flugzeug Wettbewerb* along with the J 7 with a new set of wings and balanced ailerons. The two machines were not to take part in the performance evaluations but were presented as demonstrations of the future aircraft under consideration for production. They were flown by service pilots from the Front including the redoubtable Manfred von Richthofen.

Both the 180-hp Mercedes D.IIIaü and 185-hp BMW IIIa were used in the J 10. The higher powered BMW engine proved to offer little improvement over the Mercedes at the

Above: The J 8 on the airfield, presumably for testing. Note the Morrell airspeed indicator on the starboard wing. The undercarriage has the central strut omitted but the two inverted V-struts are still in place.

low altitudes that the J 10 operated at. It was only at higher altitudes that he advantages of the BMW engine came into effect. The production CL.I arrived too late to see frontline service before the end of the war. The CL.I had the new 185-hp BMW IIIa engine. Due to the Armistice, only 43 were constructed. The CL.I served postwar with *Kampfgeschwader Sachsenberg*, the largest *Freikorp* unit, fighting in the Baltic. The Junkers CL.I proved a sturdy, effective combat aircraft whose metal structure withstood the elements better than its wood and fabric contemporaries.

Above & Below: First flown on 4 May 1918, CL.1802/18 was the production prototype for the CL.I. These photographs give a good idea of the size of the neat, two-seat fighter – an enlarged D.I. Note the revised non-balanced ailerons. The wing crosses are carried around the leading edge of the wings. There is one synchronized Spandau machine gun.

This Page: Further views of the prototype CL.1802/18. Note the stencil *"Hier anfassen"* bordered by two arrows painted on the wingtip in the photo above. This aircraft has one synchronized machine gun mounted for pilot. A Morrell airspeed indicator is mounted on top of the left wing.

Above: Dated 19 June 1918, this photograph of CL.1802/18 appears to be in the same series as those on thee preceding pages.

Above: The photograph number, 1266, and the date, 19 June 1918, ties these photographs of CL.1802/18 together. Note the radio transmitter antenna visible under the fuselage beneath the cockpits. (AHT AL0459-219)

Above: Junkers CL.I 1805/18. Two camouflage colours are visible on the upper and side surfaces. Note the two pilot's synchronized Spandau machine guns.

Above: Another view of Junkers CL.I 1802/18.

J 8 Specifications

Source	2	4
Dimensions, m		
Span	12.04	40 ft 7 in
Length	7.90	25 ft 10½ in
Height	2.66	
Wing Area	23.40 m²	253 ft²
Weights		
Empty	710 kg	1,560 lbs
Gross	1,050 kg	2,310 lbs
Fuel		
Oil		
Crew		
Payload		
Speed		
	161 km/h(max)	116 mph
Max at G/Level		
Cruising at G/ Level		
Ceiling		
Range		
Endurance		

Notes: * Competition with minimum fuel load (< 1 h)
Source:
1. Green and Scarborough.
2. Cowin, H.W. *The Junkers Monoplanes*, Profile Publications, UK, 1967.

CL.I Specifications

Source	1	1	2
Dimensions			
Span	12.15	12.00	36 ft 1¼ in
Length	7.90	7.90	25 ft 10½ in
Height	3.10	3.20	
Chord			
Track			
Wing Area	23.70 m²	23.00 m²	233 ft²
Weights			
Empty	735 kg	735 kg	1,620 lbs
Fuel	110 kg	110 kg	
Oil	15 kg	15 kg	
Crew	160 kg	160 kg	
Payload	135 kg	115 kg	
Loaded	1,155 kg	1,135 kg	2,345 lbs
Speed	190/175*	190/175*	118 mph
Range	525 km	525 km	
Endurance	3 h	3 h	
Engine	Mercedes D.IIIaü	BMW Bz IIIa	

Notes: * These sources all show the balanced, extended ailerons.
Source:
1. Wagner, W.
2. Cowin, H.W. *The Junkers Monoplanes*, Profile Publications, UK, 1967.

Above: Another view of Junkers CL.I 1802/18.

Above & Below: CL.1803/18 without any markings other than its serial. It had one sychronized machine gun for the pilot. Only six CL.I fighters were completed before the Armistice. The Junkers all metal aircraft were favourites of the pulp magazines of the 1930s, usually with misleading captions. This photograph appeared in Flying Aces magazine with the caption: "Here's the last game effort of the Germans to build an armoured trench fighter to stop the final rush of the Allies in 1918. This is the Junkers all-metal monoplane, which boasted of an armoured cockpit for trench strafing." The CL type was designed for trench strafing as one of its duties but the Junkers machine was not armoured.

Above: Completing CL.1806/18 in the Junkers factory. One D.I, three CL.I, and one J 4 fuselage can be seen in the background. The crowded conditions did not allow for ease of manufacture.

Above: This CL.I appears to have hit the port wing and then turned over. This attitude after landing seems to have been common for Junkers types and is a comment on the rough landing strips they were compelled to use. Note how the wing crosses are painted continuously over the leading edge of the wings. There is a notable contrast between the white cross outlines and the underside blue color paint. (From the August Quoos album.)

Above: An aircraft of *Kapmpfgeschwader Sachsenberg. FA* 417 on a reconnaissance mission over Kurland in March 1919.

Below: This aircraft was wrecked during a forced landing at Peterfelde in June, most probably due to fuel problems. (AHT AL0752-021)

CL.I 12921/18 "V"
***Staffel* Majewski**
***Kampfgeschwader* Sachsenberg**
Autumn 1919

CL.I "M"
***Lt.* Bruno Majewski?**
***Kampfgeschwader* Sachsenberg**
March 1919

Below: As with many ex-military aircraft after the war, a hinged cabin was cobbled together for one passenger or a small amount of cargo in what was the observer's position. Here is the Junkers modification of the J 10, complete with military crosses before the reintroduction of civil registrations. The service operated briefly in March 1919 between Dessau and Weimar, the first airline service by an all-metal aeroplane. This set of photographs were obviously posed to show how the aircraft could transport passengers and cargo. Photograph dated 12 March 1919. (STDB)

Above: August Quoos in the cockpit of CL.12608/18 that was delivered on 18 February 1919. *FA* 431 was based at Wainoden in Kurland to protect Germany's eastern borders and the German population in the Baltic states. The unit was in action against the Red Army. (From the August Quoos album.)

Below: Junkers CL.I 12619 accepted on 19 March 1919 in a slightly different camouflage scheme.

Above: Junkers CL.I in flight over Kurland (Latvia), possibly the aircraft at the top of the preceding page.

Above: Ground crew and airmen (*Ltn* Martin and *Hptm* Eberstein) pose with CL.12609/18 of *FA* 431 at Klein-Gandau (Breslau). This machine has the full serial application "Junk C.L.I. Jfa". Accepted on 20 February 1919, it appears to have an extra strut bracing the undercarriage.

Above: A pilot and observer with CL.12613/18. Note the serial application and the individual insignia. This machine has the windscreen. The weighted radio aerial fairlead may be seen under the fuselage. This machine appears to have been at Peterfelde in April 1919. (AHT AL0156-140)

Below: This CL.I in the Wainoden hangar appears to have the serial CL.12600/18. The pilot's guns are present but no windscreen. Note the LVG C.V behind. The Sachesenberg units remained at this location until March 1919, when they moved to the Vitini (Weitenfeld) airfield near Vecauce (Alt Auz), Latvia. (AHT AL072-008)

Above: 7 March 1919. *Vzfw* Herman Hackbush with CL.12600/18. The landing gear proved weak in service, and although not known for sure, this may have been the reason that some CL.I monoplanes are seen with the additional inverted V-strut undercarriage. Note the colour of the rear fuselage under the tailplane. The Zeppelin shed in the background identifies the location as Wainoden on the Kurland Peninsula (now in Latvia). (AHT AL0752-015)

Below: Same location, CL.12600/18 in background. Original caption stated this was first flight of the CL.I fighters but the undercarriage was too weak causing the pictured result. *Fw* Fritz Tödheide was associated with this crash. (AHT AL0752-012)

LVG E.I

Above: The LVG E.I was designed as a 1915 two-seat fighter design with fixed gun for the pilot and flexible gun for the gunner. The upper surfaces have been hand-painted in camouflage colors. (Peter M. Grosz Collection/SDTB)

Like so many stories in aviation history, the LVG E.I is a combination of tragedy and missed potential. A 1915 design, the E.I, serial E.600/15, was advanced for its time and perhaps the first German aircraft we would recognize as a two-seat fighter rather than an armed two-seater or battle plane.

Powered by a 120 hp Mercedes D.II, it was fitted with a flexible machine gun for the observer and a fixed, synchronized gun for the pilot. This became the standard armament for the later CL-type two-seat fighters and ground-attack aircraft. Its monoplane configuration gave the crew an excellent field of view and provided a good field of fire for the gunner. Unfortunately, the monoplane design was also the aircraft's structural weak point. The design appears to be sound, but the E.I was lost due to structural failure while enroute to the front for combat evaluation, the crew being killed. Investigation revealed that the lower wing struts had not been screwed in sufficiently and the wings collapsed. Ensuring integrity of the wing bracing before flight would seem to be an obvious and essential practice, but was either overlooked or not performed with sufficient rigor. Perhaps the design made it difficult to assess visually?

In any case, the E.I was abandoned after this fatal accident despite its promise, which seems an unfortunate over reaction to a problem that could have been readily solved with safety wire and/or better preflight inspection.

Above: The first German aircraft designed for air combat was a two-seat fighter, the LVG E.I. An advanced design for 1915, it featured a fixed, synchronized gun for the pilot in addition to a flexible gun for the gunner. It is shown here after camouflage paint was applied to the upper surfaces Tragically, the wing bracing failed during the delivery flight due to not being adequately attached and inspected, with fatal results. (Peter M. Grosz Collection/SDTB)

Above: The gunner in the LVG E.I demonstrating his excellent field of fire while the pilot visualizes shooting down an enemy aircraft with his fixed, synchronized gun, an advanced feature for 1915. (Peter M. Grosz Collection/SDTB)

Above: The LVG E.I before armament was installed and camouflage was applied. The typical LVG kinked ailerons are clearly visible. Unfortunately, no specifications survive for the E.I other than its engine type. (Peter M. Grosz Collection/SDTB)

Above: The LVG E.I cockpits. (Peter M. Grosz Collection/SDTB)

Facing Page: Two views of the LVG E.I pilot's cockpit and synchronized machine gun mechanism. (Peter M. Grosz Collection/SDTB)

Das synchronisierte MG. von Schneider
in der LVG E 20 (1915)

Wie die Synchronisierung arbeitet

Roland C.II

Above: The LFG Roland C.II was a break-through design for Roland. The exceptional streamlining for the time – it entered service in March 1916 – was developed in conjunction with wind-tunnel testing and full-size testing on a high-speed train. To reduce drag the upper wing was built directly into the fuselage without benefit of cabane struts. The C.II was the smallest, fastest German two-seater when it reached the front, and was faster than most Allied fighters. Initially armed with a single flexible gun for the observer, the second and subsequent production batches included a fixed, synchronized gun for the pilot as shown above. The speed of the C.II coupled with its effective armament made it the first German two-seater that was effective as a two-seat fighter. The relatively high-performance C.II proved a handful for many service pilots and its maneuverability and flying qualities were mediocre. The narrow gap between the wings resulted in airflow interference between them, reducing lift and increasing drag. The view upward was excellent but the view downward and forward was limited, a problem that contributed to numerous landing accidents. *Lt.* Seitz was the pilot of this C.II.

The Roland C.II was a completely new, original design by Roland. When building the Albatros B.II and C.I under license, Roland engineers became concerned about the high drag created by the multitude of bracing wires in these designs. The engineers started exploring methods to reduce drag to increase speed using the same engines currently employed. A wide field of fire for the gunner was also a design priority. This work led to the Roland *Walfisch*, or Whale, as the C.II was nicknamed in Germany.

Much of the engineering was based on the work of *Professor* Ludwig Prandtl at the Göttingen University wind tunnel. Prandtl had taught Roland chief designer *Dipl.-Ing.* Tantzen and consulted on aerodynamics for Roland. Like many significant advances in aeronautics, the Roland C.II combined new construction methods with superior aerodynamics to achieve its jump in performance. In this case, the Roland engineers used their revolutionary *Wickelrumpf* technology, which combined layers of veneer strips wrapped diagonally around a mold at an angle to each other, and glued and reinforced with fabric. This method resulted in a robust, streamlined shell that carried the structural loads. Using this new technology, the Roland C.II was designed by Tantzen, *Ober-Ing.* Cämmerer, *Ing.* Sonntag, and *Ing.* Richter.

Powered by a 160 hp Mercedes D.III engine, the Roland C.II prototype first flew on 24 or 25 October, 1915. Engine failure soon after takeoff damaged the prototype, which had to be repaired. In speed trials the C.II demonstrated a speed advantage of about 30 km/h (19 mph) over contemporary German two-seaters. The excellent speed of the Roland C.II was good enough for an initial production order for 50 aircraft, awarded 23 December 1915.

The type-tests and static load tests were performed in mid-February 1916 on airframes C.4415/15 (work #550) and C.4423/15 (work #558). During testing it was determined that the wing cellule was not strong enough, and an additional drag wire had to be installed. After that minor delay, the first five production aircraft were ready to leave the factory on 7 March 1916.

The streamlined Roland C.II arrived at the front in March 1916, where it created a sensation, and served on operations until June 1917. At the time, it was the smallest and fastest German two-seater, and certainly looked fast. Its aerodynamic shape and 160 hp Mercedes engine gave it excellent performance for the time. It was as fast as the opposing Nieuport and Sopwith Pup fighters, making it an effective combat aircraft. Because they were so much faster than the other German two-seaters, the Rolands generally operated in groups so as not to be slowed by their contemporaries.

Speed, while especially important in a combat aircraft, is not everything, and the Roland C.II's flight characteristics needed improvement. Probably the most significant shortcoming of the C.II was its limited visibility for the pilot forward and downward, making it difficult for pilots to judge their landing flair and making touch-down and landing rollout somewhat tricky and hazardous. Although landing accidents were distressingly common for WWI aircraft, the Roland C.II suffered disproportionately more than its peers.

To improve its combat effectiveness, a fixed, synchronized gun for the pilot was fitted to the C.II starting with the second production batch. The roll-over structure had to be redesigned to accommodate the addition of the gun. Additional changes to the wing structure were made midway through the third production batch to improve flight characteristics. The changes, which included moving the interplane I-struts inboard and reducing the wing span slightly, strengthened the wing. It is thought the modified version was designated the C.IIa, whose static test took place in late April or early May 1916. These aircraft reached the front in the summer.

The final significant improvement to the C.IIa was an enlarged vertical tail to improve flight stability and control. The last Roland-built production batch of 40 aircraft, ordered in June 1916, had this feature. The C.II had needed this improvement from the beginning of its service, and the change should have been made during flight-testing before the C.II went into production.

Linke-Hoffman was given an initial production order for the Roland C.IIa(Li) in July 1916 and a final production order for 40 aircraft in September. The type test took place in November 1916 and the production aircraft reached the front in early 1917. By the time these aircraft arrived at the front, new German and Allied aircraft had improved performance; the C.II had lost its performance advantage, and its operational service was nearing an end. After being withdrawn from the front, the rugged C.II served with flight schools into 1918.

Roland C.II Specifications (1st Production Version)

Engine:	160 hp Mercedes D.III	
Wing:	Span	10.33 m
	Chord	1.5 m
	Gap	1.32 m
	Area	27.96 m²
General:	Length	7.70 m
	Wheel Track	1.75 m
	Empty Weight	789 kg
	Useful Load	520 kg
	Loaded Weight	1309 kg
Maximum Speed:		165 kmh
Climb:	1000m	7 min
	2000m	14 min
	3000m	26 min

Roland C.II Production

Ordered	Qty	Serial Numbers	Designation	Remarks
Dec. 23, 1915	50	C.4413–4462/15	Rol. C.II	One MG, round rollover structure
Mar. 16, 1916	24	C.999–1023/16	Rol. C.II	Two MGs, angular rollover structure
Apr. 16, 1916	45	C.1600–1644/16	Rol. C.II	
Apr. 16, 1916	19	C.1645–1663/16	Rol. C.II/C.IIa	
Apr. 16, 1916	36	C.1664–1699/16	Rol. C.IIa	Aileron control in top wing
June 16, 1916	40	C.2700–2739/16	Rol. C.IIa	Large fin, aileron control in top wing
July 16, 1916	12	C.1851–1862/16	Rol. C.IIa(Li)	Aileron control in top wing
Sep. 16, 1916	40	C.3624–3663/16	Rol. C.IIa(Li)	Aileron control in top wing

Notes: 1. All aircraft after the first production batch had a fixed, synchronized gun for the pilot and angular rollover structures. A total of 267 production aircraft were ordered and built.
2. The production batch ordered June 16, 1916 had the large fin. All others had the original small fin.
3. The first 4 production batches used the ocarina radiator type; radiators varied on the last 4 batches.
4. The first 3 production batches had their aileron controls in the bottom wing. The last 4 batches had the aileron contol in the top wing; location of the aileron contol in the 4th batch is not known.

Above: A Roland C.II prototype rests in front of the L.F.G. factory at Adlershof. A similar photo was taken on October 24, 1915, and this photo was probably taken at the same time. An anemomenter airspeed indicator is mounted on the upper left wing for speed measurements. The aileron controls were led through the upper wing but early production aircraft ran the aileron controls through the lower wing. Although all production C.II aircraft used the 160 hp Mercedes D.III engine, at least two prototypes were tested with a 150 hp Benz Bz.III. The *Walfisch* design was patented by *Ing.* R. Richter on October 7, 1915, and the *Wickelrumpf* technology was a closely-held industrial secret until 1918 when publicity photos of the Pfalz factory revealed some details of the process.

Above: The first Roland C.II series production aircraft with company test pilot *Hptm. Freiherr* von Thüna in the cockpit on February 4, 1916. The fin and rudder were slightly increased in area compared to the prototypes, but needed to be even larger for better flying qualities, something finally addressed in the last production batch ordered from LFG. No rollover cage is fitted to this early production aircraft.

Above: The Roland C.II arrived at the front when the comparatively primitive Fokker *Eindeckers* were still the leading German fighters. This early production C.II lacked a fixed, synchronized gun for the pilot so was not yet a two-seat fighter, but it was faster than the *Eindeckers*. There is a great contrast in production technology between the two aircraft.

Below: This early-production Roland C.II shows the clean lines of this break-through design. Eyes and mouth have been painted on to emphasize the *Walfisch* design's resemblence to a fish. Unlike the prototypes, the early production aircraft had their aileron cables routed through the lower wing, then rising upward externally to move the ailerons. The streamlined 'ocarina' exhaust was a distinctive feature of many C.II aircraft.

Above: Roland C.II of *Feldflieger-Abteilung* (A) 227 surrounded by the ground crew. *Lt.* Stuhldreher sits on the top wing and *Lt.* Wilhelm Allmenröder sits on a wheel.

Below: Later production C.II with fixed gun for the pilot and bomb racks under the center section.

Right: Four mid-production C.II aircraft dismantled for transportation. The fixed machine gun for the pilot was fitted to all but the first production batch and required the rollover structure to be redesigned. The ear radiators, clearly visible here, detracted from the streamlined fuselage lines.

Above: A mid-production C.II in the snow displays its dark camouflage. Unfortunately, the unit and crewmen are unknown.

Right: This closeup of a camouflaged, mid-production C.II shows the crew and armament. The observer has signal flares mounted in a rack aft of his window and the pilot has a rear-view mirror mounted on the turnover pylon.

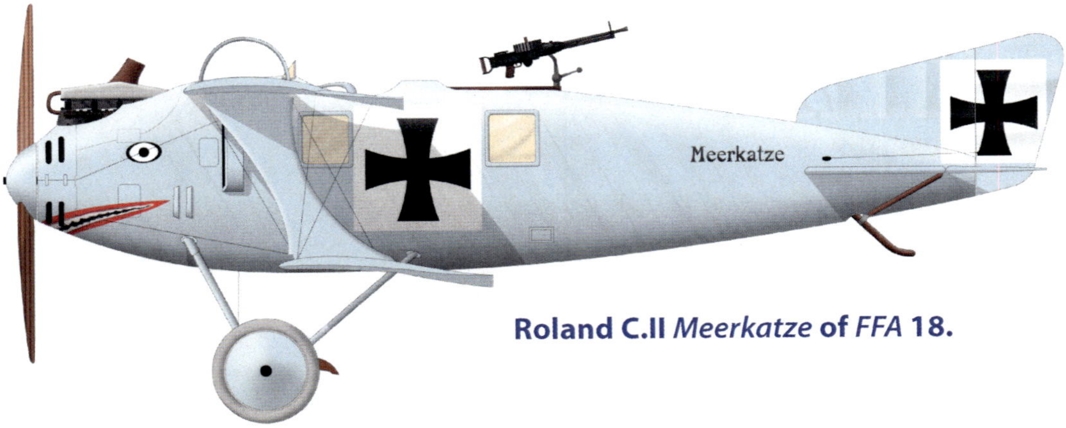

Roland C.II *Meerkatze* of *FFA* 18.

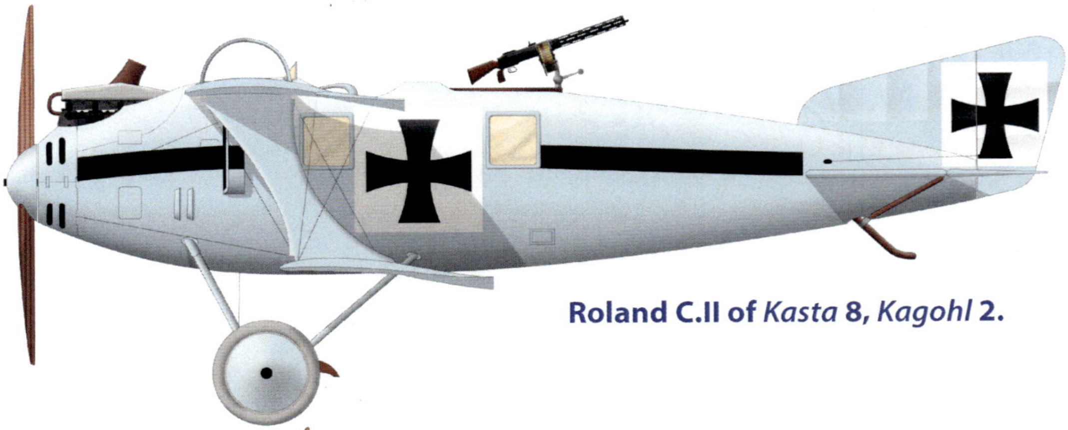

Roland C.II of *Kasta* 8, *Kagohl* 2.

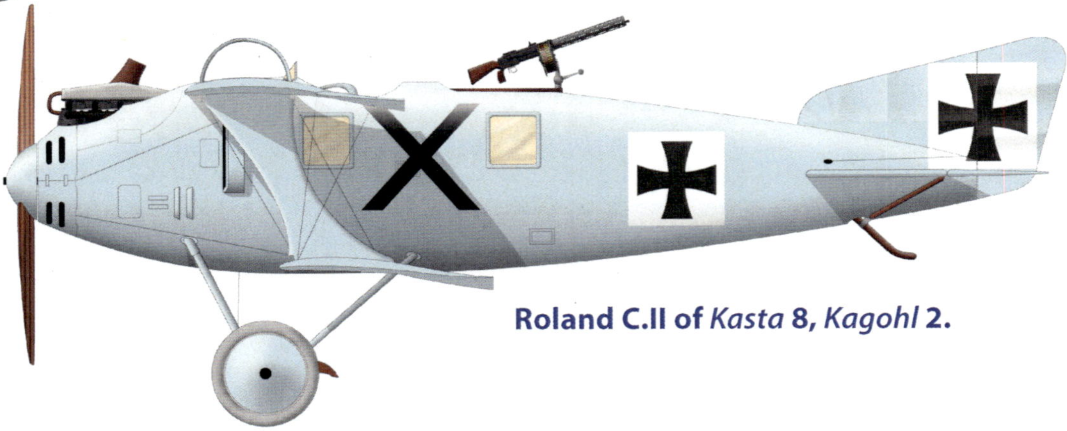

Roland C.II of *Kasta* 8, *Kagohl* 2.

Roland C.II of *Kasta* 8, *Kagohl* 2. This aircraft may have been flown by Manfred von Richthofen because he was in that unit when this aircraft served there.

Roland C.II of *Schusta* 27.

Roland C.II flown by *Hptm*. Eduard *Ritter* von Schleich of *Schusta* 28 in original configuration.

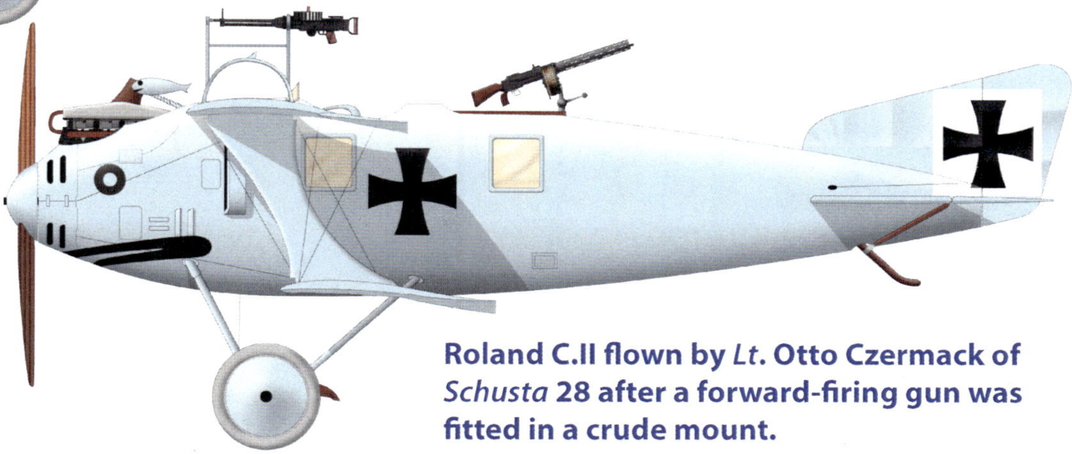

Roland C.II flown by *Lt*. Otto Czermack of *Schusta* 28 after a forward-firing gun was fitted in a crude mount.

Roland C.II with shark mouth of *Flieger-Abteilung* 62.

Above: This closeup of the mid-production C.II shown in full on page 17 shows the armament details. The observer's window has been modified to open by sliding it aft. *Lt.* Seitz was the pilot.

Above: The aircrew of this mid-production C.II have retained their sense of humor at the front and decorated their aircraft to suit themselves.

Above: A camouflaged Roland C.II of *Flieger-Abteilung* 62 with tactical number '7' painted on its fin and a shark mouth.

Above: A Roland C.II of *Flieger-Abteilung* 62 with shark mouth is photographed with its flight crew in the center (*Oblt.* Maximilian von Cossel at left, and to his left, *Vzfw*. Rudolf Windisch) flanked by its ground crew on the Eastern Front.

Roland C.II of *Lt*. Rudolf Windisch of *Feldflieger-Abteilung* 62; it has unit-applied colors painted on the fuselage, upper surfaces, and interplane struts for camouflage.

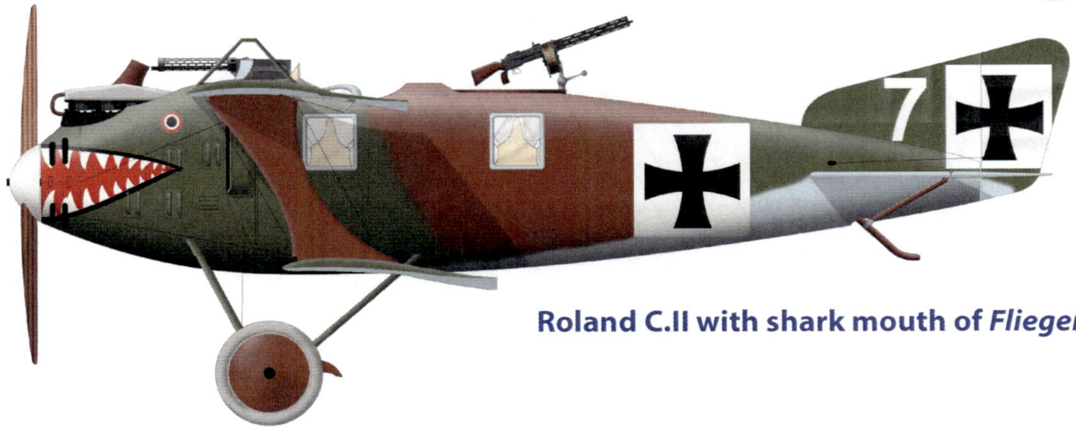

Roland C.II with shark mouth of *Flieger-Abteilung* 62.

Roland C.IIa with large tail, *Schusta* 27.

Roland C.IIa with large tail of an unknown unit.

Above: Roland C.IIa from the final Roland production batch with the large fin needed for greater stability. It is camouflaged and the pilot's machine gun has been raised well above the normal position.

Above: Large-tail Roland C.IIa 2701/16 of *Feldflieger-Abteilung (A)* 292b in the spring of 1917 at Houplin airfield in northern France. It has had the aft fuselage repaired; external stringers for additional strength have been added to the aft fuselage. The pilot is *Lt*. Otto Fuchs, who later became a fighter pilot and served with *Jastas* 30, 35b, and 77b. The observer is his brother *Lt.d.R.* Rudolf Fuchs, who later became a fighter pilot who served with *Jastas* 77b and 35b.

Right: Roland C.II with *Lt*. Boes in the observer's cockpit and *Oblt*. Dostler at right. *Lt*. Boes is complaining to the *Waffenmeister* Lademacher at left about gun stoppage problems.

Roland C.III

The Roland C.III was a development of the C.II with conventional two-bay interplane struts replacing the streamlined, single-bay I-struts of the C.II. While the I-struts aided streamlining, they also obscured the view to the sides and the conventional struts improved the view for the crew. The C.III also was fitted with the more powerful 200 hp Benz Bz.IV engine typically used in second-generation C-types like the DFW C.V, etc. The C.III remained a single prototype and was destroyed when LFG's Adlershof factory burned down on 6 September 1916.

Roland C.III Specifications	
Engine:	200 hp Benz Bz.IV

Above & Below: Clearly derived from the C.II, the Roland C.III had larger wings with conventional two-bay interplane bracing. Intended for the more powerful 200 hp Benz Bz.IV engine, it appears like a 160 hp Mercedes D.III is installed here, perhaps for initial flight testing. Armament was the standard fixed Spandau for the pilot and flexible Parabellum for the observer. The sole prototype was destroyed when LFG's Adlershof factory burned.

Roland C.IV

There is no known information for the Roland C.IV. Because the Roland-built Albatros C.X was produced about the same time *Iflieg* rationalized the aircraft designation system, the Albatros C.X(Rol) could originally have been designated Roland C.IV, C.VI, or C.VII in company records. In any case, apparently there is no original Roland design built that was designated Roland C.IV.

Later Roland produced other two-seaters under license, the Hannover CL.II(Rol) and Halberstadt CL.IV(Rol). These were too late to receive a confusing designation under the old system, but perhaps Roland management assigned a Roland designation to them, or, more likely, these remained unbuilt Roland projects.

Roland C.V

The Roland C.V was a compact, streamlined two-seater derived from the Roland D.II fighter. Like the D.II the C.V was a single-bay biplane that utilized *Wickelfumpf* technology for the fuselage structure. The C.V was powered by a 160 hp Mercedes D.III engine and had one fixed and one flexible machine gun. Apparently there was no significant advantage over the C.II and only one C.V was built.

Roland C.V Specifications	
Engine:	160 hp Mercedes D.III

Above: The Roland C.V was derived from the D.II fighter. It had the same 160 hp Mercedes D.III engine used in the D.II and larger wings with conventional single-bay interplane struts. Armament was the standard fixed Spandau for the pilot and flexible Parabellum for the observer. Only one Roland C.V was built.

Rumpler 6A2

Above: The Rumpler 6A2 first flown in January 1916 was a two-seat fighter. The first version used 'X'-struts. (The Peter M. Bowers Collection/The Museum of Flight)

The single-bay 6A2 was designed as a two-seat fighter. It was a refined, single-bay derivative of the Rumpler C.I reconnaissance airplane. It remained a prototype.

Powered by a 160 hp Mercedes D.III, its first test flight in January 1916 revealed problematic handling qualities, but extensive experimentation and modification finally gave the 6A2 'excellent flight characteristics' according to Friedrich Budig.

Unfortunately, climb rate was less than the C.I reconnaissance airplane and the wings were too 'elastic', so the X-struts were replaced with parallel struts. The 220 hp Mercedes D.IV straight-eight engine was also tried and gave very good climb rate but vibration was excessive. Records are ambiguous about whether the 160 hp Mercedes D.III was re-installed. Both flight testing and calculations indicated the 6A2 would not meet its requirements, and it was abandoned in favor of the enlarged, two-bay 6A5 prototype that became the C.III of otherwise similar appearance. The intermediate 6A3 and 6A4 were designed but probably not built.

Rumpler 6A2 Specifications

Engine:	160 hp Mercedes D.III	
Wing:	Span	10.20 m
	Area	36.00 m^2
General:	Empty Weight	780 kg
	Loaded Weight	1260 kg
Climb:	1000m	6.0 min
	2000m	16.5 min
	3000m	30.7 min
	4000m	69.0 min
Ceiling:		4200m

Above & Below: The Rumpler 6A2 was powered by a 160 hp Mercedes D.III. The first version shown here used 'X'-struts.

Above: Rumpler test pilot Gustav Basser at left in front of the Rumpler 6A2 at Johannisthal. Although not a successful design, the 6A2 was instrumental in development of the larger, two-bay 6A5 that became the Rumpler C.III and led to the Rumpler C.IV. The 6A2 introduced the basic fuselage and tail assembly and even the wing planform that are seen in the later and larger, more powerful two-bay Rumpler C.III reconnaissance airplane.

Above: The second version of the 6A2 used parallel struts for greater torsional strength of the wing cellule, giving improved aileron response and better controllability.

Zeppelin-Lindau C.I / CL.I

Above: The ZWL C.I, aircraft number 2, on the airfield in Zech before it was transported to Adlershof. The five-color camouflage from the wings was transferred to the fuselage as well. (Airbus Group)

Ferdinand Graf Zeppelin was born on July 8, 1838 in Constance on Lake Constance. A German army officer, he was a military observer to the American Civil War. In this capacity he flew in a Union Army observation balloon on August 19, 1863. This showed him the potential in lighter than air aviation and in 1894 he presented a concept for a rigid airship to a council of experts. Despite finding no support, he persisted and it 1898 he founded the "Gesellschaft zur Förderung der Luftschiffahrt" (Society for the Promotion of Aerostation). More than half of the capital stock of came from the count's private assets. Construction of the first airship, LZ 1, started in 1899 and its first flight was on July 2, 1900, in Friedrichshafen.

The parent company of all subsequent Zeppelin companies was "Luftschiffbau Zeppelin GmbH", which had initially been founded by Count Zeppelin exclusively for the construction of rigid airships.

In August 1914, at the beginning of the war, he gave two simultaneous orders. To his engineer Claude Dornier from Friedrichshafen Zeppelin gave the order to build huge metal seaplanes. This became the Zeppelin-Linau company.

To Professor Baumann from Stuttgart Zeppelin gave the order to build huge land planes. This became the Zeppelin-Staaken company. Initially housed at Gotha, the potential problem of air raids due to proximity to the boarder forced relocation to Staaken near Berlin.

Around this same time, a new subsidiary of Luftschiffbau, Flugzeugbau-Friedrichshafen GmbH, was founded for the construction of seaplanes. Zeppelin Luftschiffbau left its workshops in Manzell to it as a factory.

Thus Zeppelin, whose name is indelibly associated with the giant airships he created, founded three new airplane companies at the beginning of WWI. Although the Zeppelin-Lindau company was founded to build giant metal seaplanes for the Navy, only Rs.III was delivered to the Navy during the war for operations due to the severe difficulty involved in developing this new technology.

Zeppelin-Lindau later expanded its business to build metal airplanes for the Army. The Army bought 20 times the number of airplanes than the Navy, so this was an obvious decision.

The first two-seat combat aircraft designed by Zeppelin-Lindau was the C.I, ambiguously also known as the CL.I due to its design. The C.I/CL.I was a single-bay biplane of conventional configuration but all-metal construction. The C.I/CL.I was a small aircraft designed for speed; it was a relatively small aircraft with a single-bay of bracing and was powered by a Mercedes D.III/D.IIIa engine. Armament was

Above & Below: After the take-off issues on November 3rd, 1917, a larger rudder was installed on the C.I. The aircraft is shown after camouflage fabric was applied to the airframe. (Airbus Group).

standard for C/CL types, a fixed, synchronized gun for the pilot and a flexible gun for the observer.

While the configuration was entirely conventional, the structural design and construction was revolutionary. Not only was the structure all-metal, the design was stressed-skin, semi-monocoque. A monocoque structure, like an egg, is one in which the shell supports all the structual loads. A semi-monocoque struture is a combination of the skin and supporting structure (ribs and spars) supporting the loads.

This was a revolutionary advance, and this basic design approach is still used for new airplanes today. And it had many challenges. First, a satisfactory light alloy had to be developed; it had to be light, strong, and malleable. Many alloys are too brittle to be suitable.

Second, new design techniques had to be developed for this much more sophisticated structure. Third, more detailed technical drawings and specifications were required for metal structures compared to wood structures. Fourth, new forming techniques had to be developed to make the parts, and new fasteners had to be designed and fabricated.

The net result was that metal aircraft took a lot more time and work to design and build than wood airplanes. The design had been proposed in October 1916 and the maiden flight of the C.I/CL.I was attempted on November 3, 1917. Due to the undersize rudder, the aircraft turned 90° to the right upon application of takeoff power and the pilot aborted the take-off. After a larger rudder was designed and fitted, the actual first flight took place on 14 November.

Above, Below, & Facing Page, Above: The Zeppelin C.I prototype at Adlershof before camouflage fabric was applied to the fuselage. (PM Grosz collection/STDB)

Continued problems with the swing on take-off was solved by lengthening the rear fuselage. Test flying revealed issues with the engine, which had to be replaced twice before satisfactory results were achieved. Finally, on 29 November a speed to 190 km/h was measured during a test flight.

Idflieg's order for three prototypes was received on December 8, 1917, and the serials 15800–15802/17 were assigned. The second and third airframes were completed by the end of 1917 and were shipped by train to Adlershof in early 1918. The C.I acceptance flights at Adlershof took place on January 10. Minor modifications were then made. The radiator, mounted in the center of the upper wing, was not satisfactory and needed to be moved or replaced. However, due to the metal construction, this could only be done on the C.II/CL.II. During the test, Idflieg conducted stress tests and created a specification sheet where the type was designated the CL.I, reflecting its actual design role.

Zeppelin-Lindau CL.II

Above: The upper side of the wings as well as the horizontal tail surfaces of the Cl.II were covered in the tri-color camouflage fabric of the navy. The fixed elements of the empennage, on the other hand, were covered with the five-colored camouflage fabric of the army.

Unfortunately, one of the Zeppelin-Lindau CL.I prototypes crashed fatally on March 9. The cause could not be determined; the pilot, Franz Küstner, was killed and the wreck was totally destroyed. Some strengthening of certain parts of the airframe was recommended; these changes and some others requested after test flying the CL.I were incorporated in the CL.Ia, later renamed the CL.II.

Most of these changes were not readily visible form the exterior of the CL.II; the one distinctive change that was clearly visible was the use of a nose radiator to eliminate the concerns about the radiator mounted in the middle of the upper wing of the CL.I. The span of the upper wing was increased by 10 cm to improve climb rate, the entire tail was re-enginerred, the engine was the same as the CL.I, and an additional fixed gun was added for the pilot.

Production of the CL.II was delayed to the summer of 1918 due to the higher priority of other projects at the Zeppelin-Landau factory, specifically the new Rs.IV giant flying boat, the D.I single-seat fighter, and the Zeppelin (Ja) C.II. Factory photos show 11 completed CL.II fuselages plus four others under construction. The maiden flight of the new CL.II was made on August 17, 1918, where a top speed of 164 km/h was reached.

Likely, due to delays and bureaucracy, just two CL.II aircraft were completed by the Armistice. It is not known why Idflieg did not strongly suppor introduction of the CL.II and the all-metal D.I single-seat fighter. Both types offered excellent speed and climb rate and their metal construction made them much more durable than the competing types made of wood and fabric. These types were much more susceptible to weather, a particular problem with the frequent relocation of aerodromes due to the fluid situation on the ground. While it was difficult to keep wood and fabric aircraft in combat condition for more than a few months, the Zeppelin C.II (not CL.II) aircraft supplied to Switzerland were used from 1919 to 1927.

Above: Zeppelin-Lindau Cl.II aircraft number 2 on the airfield in Zech before its maiden flight. With its integrated nose radiator it was a very state of the art aircraft for the time. (Airbus Group)

Above: The Cl.II aircraft number 2 on the airfield in Zech before its maiden flight. (Airbus Group)

Above & Below: The Cl.II aircraft number 2 on the airfield in Zech before its maiden flight. Tooling marks on the fuselage metal are clearly visible. (Airbus Group)

Made in the USA
Middletown, DE
26 May 2024

54819741R00093